Principles of Composite
Material Mechanics

Principles of Composite Material Mechanics

Contributors

Shilko Serge et al.

AURIS
Reference

www.aurisreference.com

Principles of Composite Material Mechanics

Contributors: Shilko Serge et al.

Published by Auris Reference Limited
www.aurisreference.com

United Kingdom

Principles of Composite Material Mechanics

ISBN: 978-1-78154-945-2

British Library Cataloguing in Publication Data
A CIP record for this book is available from the British Library

Printed in the United Kingdom

Exclusively distributed by CBS Publishers & Distributors Pvt. Ltd.

Sales & Distribution Rights only for India, Pakistan, Bangladesh, Sri Lanka, Nepal and Bhutan.This book is not to be sold outside these territories.

Contents

List of Abbreviations

ACM	Adaptive composite materials
AFM	Atomic force microscope
BSE	Backscattered
CF	Carbon fibers
CEMHYD3D	Cement Hydration and Microstructure Development Model
cov	Coefficient of variation
DM	Digital Microscopy
DFB	Distributed feedback
EMT	Effective Medium Theory
EMC	Electromagnetic compatibility
EMI	Electromagnetic immunity
FEM	Finite element models
GNs	Graphene nano-sheets
DE	Diffraction efficiency
ITO	Indium tin oxide
MG	Maxwell Garnet equation
MMC	Metal-matrix composite
NIST	National Institute of Standards and Technologies
NLC	Nematic liquid crystal
OM	Optical Microscopy
PE	Polyethylene
PDD	Powder deposition density
PID	Proportional–integral–derivative
RF	Radio-frequency
R&D	Research and development
SEM	Scanning electronic microscope
SOAP	Simple Object Access Protocol
SMEs	Small and Medium Enterprises
WP	Work piece

List of Contributors

Shilko Serge
V.A. Belyi Metal-Polymer Research Institute of NASB, Belarus

E. Dado
Netherlands Defence Academy, Breda, The Netherlands

E.A.B. Koenders
Delft University of Technology, Delft, The Netherlands, COPPE-UFRJ, Programa de Engenharia Civil, Rio de Janeiro, Brazil

D.B.F. Carvalho
(PUC-Rio), Rio de Janeiro, Brazil

Konstantin N.
Institute for Theoretical and Applied Electromagnetics, Russian Academy of Sci., Moscow, Russia

Marina Y. Koledintseva
Missouri University of Science and Technology, Rolla, MO, USA

Eugene P. Yelsukov
Physical and Technical Institute, Ural Branch of Russian Academy of Sci., Izhevsk, Russia

R. Caputo
LICRYL (Liquid Crystals Laboratory, IPCF-CNR), Center of Excellence (CEMIF. CAL) and Department of Physics, University of Calabria, Arcavacata di Rende, 87036 Cosenza, Italy

L. De Sio
LICRYL (Liquid Crystals Laboratory, IPCF-CNR), Center of Excellence (CEMIF. CAL) and Department of Physics, University of Calabria, Arcavacata di Rende, 87036 Cosenza, Italy

A. Veltri
LICRYL (Liquid Crystals Laboratory, IPCF-CNR), Center of Excellence (CEMIF. CAL) and Department of Physics, University of Calabria, Arcavacata di Rende, 87036 Cosenza, Italy

A. V. Sukhov
Institute for Problems in Mechanics, Russian Academy of Science, Moscow 119526, Russia

N. V. Tabiryan
Beam Engineering for Advanced Measurements Company, Winter Park, Florida 32789, USA

C. P. Umeton
LICRYL (Liquid Crystals Laboratory, IPCF-CNR), Center of Excellence (CEMIF. CAL) and Department of Physics, University of Calabria, Arcavacata di Rende, 87036 Cosenza, Italy

S. Paciornik
DEMa PUC-Rio, Rio de Janeiro, Brazil

J. d'Almeida
DEMa PUC-Rio, Rio de Janeiro, Brazil

Yong X. Gan
Department of Mechanical, Industrial and Manufacturing Engineering, College of Engineering, University of Toledo, 2801 W Bancroft Street, Toledo, OH 43606, USA

B.L. Sharma
Department of Chemistry, University of Jammu, India

Parshotam Lal
Department of Chemistry, University of Jammu, India

Ali Emamian
University of Waterloo, Canada

Stephen F. Corbin
University of Waterloo, Canada

Amir Khajepour
University of Waterloo, Canada

Chien-Lin Huang
Department of Fiber and Composite Materials, Feng Chia University, Taichung City 40724, Taiwan

Ching-Wen Lou
Institute of Biomedical Engineering and Materials Science, Central Taiwan University of Science and Technology, Taichung City 40601, Taiwan

Chi-Fan Liu
Office of Physical Education and Sports Affairs, Feng Chia University, Taichung 407, Taiwan

Chen-Hung Huang
Department of Aerospace and Systems Engineering, Feng Chia University, Taichung City 40724, Taiwan

Xiao-Min Song
Laboratory of Fiber Application and Manufacturing, Department of Fiber and Composite Materials, Feng Chia University, Taichung City 40724, Taiwan

Jia-Horng Lin
Laboratory of Fiber Application and Manufacturing, Department of Fiber and Composite Materials, Feng Chia University, Taichung City 40724, Taiwan
School of Chinese Medicine, China Medical University, Taichung City 40402, Taiwan
Department of Fashion Design, Asia University, Taichung City 41354, Taiwan

Zachary Zachariev
Institute of Polymers, Bulgarian Academy of Sciences, Bulgaria

Preface

A composite material is a material made from two or more constituent materials with significantly different physical or chemical properties that, when combined, produce a material with characteristics different from the individual components. The text *Principles of Composite Material Mechanics* presents a unique blend of classical and contemporary mechanics of composites technologies. Bionics principles, abnormal elasticity, and moving interfaces of adaptive composite materials have been focused in first chapter. Second chapter describes the concept of so-called virtual laboratories that are based on a netcentric approach. Third chapter discusses frequency dependences of effective material parameters (permittivity and permeability) of different types of composites. Features and applications of policryps composite materials have been described in fourth chapter. Fifth chapter focuses on digital microscopy and image analysis applied to composite materials characterization. Sixth chapter deals with the effect of interface structures on the mechanical properties of fiber reinforced composite materials. Seventh chapter authenticates with experimentally investigated strength data that the physical properties, and in particular, the mechanical behavior of a material depend on the growth habits to produce the modal microstructure of the material. Eighth chapter discusses the development of an *in-situ* laser cladding technique to deposit a TiC-based MMC coating on a steel substrate with no cracks and excellent bonding to the substrate and a high hardness. Ninth chapter aims to examine the properties of composites that different carbon materials with different measurements can reinforce. Last chapter deals with new superhard ternary borides in composite materials.

Chapter 1

ADAPTIVE COMPOSITE MATERIALS: BIONICS PRINCIPLES, ABNORMAL ELASTICITY, MOVING INTERFACES

Shilko Serge

V.A. Belyi Metal-Polymer Research Institute of NASB, Belarus

INTRODUCTION

Requirements imposed on artificial materials are constantly rising with time. Along with lately requisite properties, including stability of physical and mechanical characteristics, linearity of the equation of state and unambiguity of response to disturbance, there arose a problem of a complex active response to varying outer conditions. In other words, a tendency is observed of increasing number of material functions acquiring the features of intellectual systems. So, obvious prototypes of these materials turn to be biosystems, from the one hand, and computer monitored technical systems able to reproduce intellectual behavior using sensor, processor and executive functions (including effector function and response action), from the other hand, plus feedforward and feedback. Although means of these properties realization can't be similar in artificial materials and above mentioned natural prototypes, generalizations obtained at the junction of the materials science, bionics and cybernetics allow to formulate the conceptual principles and to consider probable ways of the named interdisciplinary problem solution.

Recent reviews and terminological discussions in the field have confirmed actuality of the structural and functional analyses of smart composites, including functional nanomaterials [Bergman & Inan, 2004]. However papers, devoted to such materials (e.g., self-controlled membranes on hydrogel base [Galaev, 1995]) are commonly reduced to creation of sensors and actuators. Less attention have received principles and models of adaptive reactions in composites. The adaptive mode of reinforcing and self-assembling in smart materials [Schwartz, 2007] has been studied below in the form of phenomena caused unusual elastic properties of auxetic and multimodule materials. The development of adaptive composites allows us to hamper the failure process and promotes reliability and service life of products for different technical applications.

ADAPTIVE COMPOSITES IN CLASSIFICATION OF MATERIALS

Classification of Materials

The first stage of the present study is classification of materials with account of interrelations found between structure and functions as well as analysis and modeling of a subclass of intellectual systems, namely adaptive composite materials (ACM). Some of the assumptions put forward by the authors are based on the theory of functional systems and synergism [Prigogine & Stengers, 1984]. Three generations of materials which can be discriminated in the proposed classification, are given in Table 1.

Table 1. Evolution of structure and properties of materials

Generation of materials	Structural-and-functional characteristics	Means of property regulation	Factor determining optimum result
Traditional material	Monofunctional single-component material	Properties determined a priori by the origin of component material	Initial property of monocomponent
Composite material	Monofunctional polycomponental material with fixed boundaries between components	Properties are efficiently regulated technologically based on principles of additivity and synergism	Initial property of components and intermediate layers
Smart (adaptive) composites	Polyfunctional polycomponental material with movable boundaries between components	Self-regulation of structure based on sensor, processor and effector functions and feedforward and feedback channels	Efficiency of sensing extreme effects and elimination of refusals

The first generation is traditional materials including monofunctional medium whose properties are determined by the nature and initial quality of a single component. The next are traditional composites with a prominent structural hierarchy, being also monofunctional. They are characterized by stability of inner and external boundaries, i.e. fixed structure of components, intermediate layers and the composite as a whole. Adaptive materials with coordinated functions and active behavior belong to the third advanced generation of materials. These systems perceive outer effects at unchanged function owing to, presumably, structural self-organization. In this connection, the mobility of the component boundaries should be remembered as an indispensable property of smart materials, which is not present in traditional composites.

The qualitative transition of materials from the passive to active functioning is shown in Table 2. Naturally, prerequisites of such a transition are formed at the levels of two preceding generations. Thus, transformation of one physical field into another (e.g., piezo- or photo effects) is probable at the stage of monofunctional material. The creation of qualitatively new (emerged), including forecast properties, is a logical continuation of the additive and synergetic principles of composite production. This precedes the development of adaptive composites, being a subclass of smart systems with the dominating adaptive strategy.

The suggested classification makes it possible to forecast other unknown materials of the intellectual type, for example, capable of self-destruction "kamikaze", those ensuring partial or full restoration "regenerators" and materials offering programmed control of the environment ("cyber") and implicit ("incognito") ones. These subclasses constitute a new type of "ecophilous" materials which behavior supports homeostasis of the environment.

Table 2: Systematization of materials by general criteria

Functional evolution	Degree of activity	Degree of intellect	Functioning quality	Mode of behavior
mono-functional	passive	"trivial"	material	"predictable"
	active	"wit" (functional)		
poly-functional	active	smart (adaptive)	material = part	"indefinite"
				"egoist"
			material = system	"time-server"
		"wise" (ecophilous)	material = medium	"kamikaze"
				"regenerate"
				"cyber"
				"incognito"

Adaptive Composites

Relative simplicity of ACM is due to their orientation aimed to fulfill only the adaptive function of the part or a system in contrast to a higher status of the material-medium subclass (Table 2). However the adaptive composite is formed rather in time than by a mechanical mixing of structural components, and evolutionizes as a specific unit by coordinating interrelated physical processes based on an imparted optimum criterion. In this case, the emergence of macrostructure is specified by origination of collective modes under the action of fluctuations, there competing and, finally, by selection of the most accommodated mode or their combination [Prigogine & Stengers, 1984]. The structures themselves could be described in physical terms as types of adaptation to outer conditions.

Self-organization of Material Structure

Reaction of a material due to mutual coordination of structural and functional parameters of microsystems characterizes it as an open self-regulating system. Selection of the mode of behavior in response to outer effect does not arise from the principle of the least action, neither from the principle of compulsion (Gauss principle) nor from that of the utmost probability. Active response systems eliminate (or subordinate) contingency. This makes grounds to speak about a programmed behavior of the system, i.e. the decision is made according to the inner criteria determined by the structure itself and system parameters, which substantiates the necessity of direct and reverse connection channels.

It follows from the above said that to form a more complex processor function of ACM it is possible to use the universe phenomenon of self-organization, which is not limited to only systems of higher organization and functional complexity and isn't a monopoly of bio- or social systems. A self-organizing system is understood as a system capable of stabilizing parameters under varying outer conditions through directed ordering of its structural and functional relations aimed at withstanding entropic factors of the environment, which helps to preserve its characteristics as an integral formation [Prigogine & Stengers, 1984]. The material formed by combining its components acquires the characteristics of a composite structure, which is a notion nonequivalent to the structure of its constituents. This fact raises composite materials to a higher structural level and admits the probability of per layer differentiation of the functions in order to reach the integral control system. In our view, to realize adaptation mechanism to outer conditions in composite materials, it's worthwhile considering the combination of different scale physical processes, where we single out at least 4 structural levels: molecular, mesoscopic, macroscopic and polycomponental (Figure 1).

The molecular level is the basic one at programming material behavior. This is because its scale in polymer composites corresponds to cooperative effects of segmental mobility and conformal rebuilding that provide conditions for self-organization in high-molecular bodies. Just here the processor function is realized as a capacity for estimating variations due to outer effects and as a tool formulating the character and force of response based on stationary characteristics of the microsystem. Also, the effector function is fulfilled here for exciting reverse reactions by varying characteristics of the microsystem on a self-organization base. The mesoscopic level performs the sensor function as an ability to perceive outer effects. Non-equilibrium processes are initiated at this level changing molecular structure and supporting the interaction of direct and reverse channels between the levels. The macroscopic level makes provision for the mobile function as a reorganization of the initial subsystems (components) aimed at preserving the behavior model. The mobile function is also realized at the polycomponental level, though intention in this case to provide the system (material = article) functioning as a whole. To organize control, the processes relating to the mention levels should be coordinated using functional links between them.

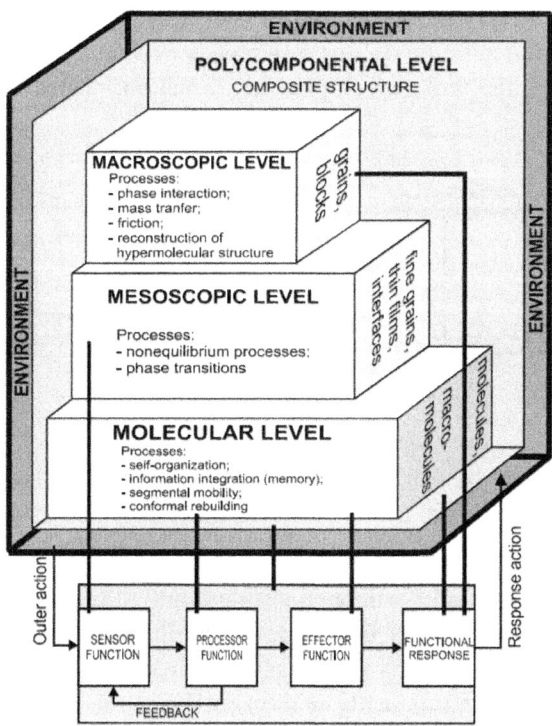

Figure 1: Differentiation of structural levels at ACM development

It is to be remembered that polymer composites are potential carriers of intellectual properties. Namely, they are sensitive to physical fields, i.e. show a sensor function; make it possible to carry out the actuator function (shape memory of thermosetting resins, etc) and, finally, among all other artificial material media they most closely approach the living nature (biotissues are usually built of high-molecular compounds). The study of synergetic phenomena in nonliving nature as a linking element between analogous processes in original objects will, in our opinion, provide a possibility to find structural-and-functional bioprototypes of adaptive composites.

SELF-REINFORCING IN AUXETIC COMPOSITES

The effort towards improving the performance of novel devices based upon realisation of non-linear and non-trivial (anomalous) deformation properties of materials is the aim of many current investigations. First, we shall consider the materials with a negative Poisson's ratio, v, termed 'auxetics'. Data structuring, examination of the mechanisms of generating the negative Poisson's ratio and analysis of likely applications for auxetics have been discussed recently in [Wojciechowski et al, 2007] and reviewed particularly in [Koniok et al, 2004]. Poisson's ratio affects a very important mechanical property, i.e. compressibility of a material. Under a uniaxial stress, auxetics expand/contract at the direction perpendicular to the tension/compression direction, respectively as shown in Figure 2.

(a)   (b)

Figure 2: The deformation mode of an auxetic material under uniaxial stress: (a) compression, (b) tension. Initial configuration before loading has been shown by dashed lines.

This property should influence stiffness and slip under contact loading, and in this way allow control over deformability and friction characteristics of composites and joints based on auxetics. As will be shown, the contact characteristics vary dramatically with variation of the sign of Poisson's ratio. In the classical elasticity theory for isotropic bodies [Landau & Lifshitz, 1986] Poisson's ratio $v = (3K - 2\mu) / (6K + 2\mu)$, where μ, K are the shear and volume moduli respectively, the Poisson's ratio of isotropic bodies can vary in the limits $-\leq\leq$ 1 0 v .5 . The upper limit corresponds to incompressible materials, e.g. rubber, whose volume remains constant at significant shape variations, the lower one belongs to the materials preserving their geometrical form with changing volume.

Several natural and artificial auxetic materials have been described to date, but experimental and theoretical studies of the adaptive frictional and mechanical properties of these materials are not still well developed [Baughman & Galvao, 1993]. For example, there exists the possibility for realisation of self-reinforcing or self-locking effect in contact joints containing auxetic components. As a result, this effect would bring about a significant increase in the bearing capacity of frictional joints or shear strength of the fibre – matrix interface under mechanical or thermo-mechanical load. Of specific interest here is the study of the self-locking effect under contact deformation of anisotropic auxetics based on directionally reinforced composites. This is because such materials may posses Poisson's ratios of much less than -1 (v < -50) and considerable strength due to their directional reinforcement.

The approaches available for creation of composites with v < 0 assume either the use of individual auxetic components or formation of an auxetic composite – a combination of structural units of mesoscopic level (pores, granules, permolecular formations of polymers, etc). To study friction effects under contact loading the existing estimates of elastic properties of quasi-isotropic and anisotropic composites should be taken into account.

Auxetic Inclusions (Quasi-Isotropic Auxetics)

In [Wei & Edvards, 1999] the mechanical characteristics of a composite with ellipsoidal and spherical particles were calculated for the case of randomly distributed filler particles. Simulation results under different ratios of filler stiffness to matrix stiffness, for 45% volume fraction, are presented in Table 3.

Table 3: Effective Poisson's ratio vc of the composite at v = 0

Inclusion geometry	0.1	1.0	10
Disc (2D)	-0.3020	-0.2856	0.1216
Disc (3D)	-0.0385	-0.3575	-0.7387
Sphere	-0.0624	-0.2081	0.0650
Wedge (2D)	-0.2679	-0.2266	-0.0508
Needle (3D)	-0.0555	-0.1714	-0.0562

The possibility of obtaining auxetic composites using filled polymers has been considered in [Kolupaev et al, 1996]. The authors have obtained such composites using thermoplastic polyurethane with ultra-dispersed (0.3-1 μm) particles of tungsten, iron and molybdenum having $v \approx$ -0.2-0.4. The composite possessed auxetic properties due to internal stresses σ_{in} produced by the inclusions in the matrix in the range 0.97 MPa $< \sigma_{in} <$ 7.11 MPa.

Non-auxetic Inclusions (Anisotropic Auxetics)

Let us consider a composite formed by the oblique packing of fibres in an elastic incompressible elastomeric matrix (Figure 3a).

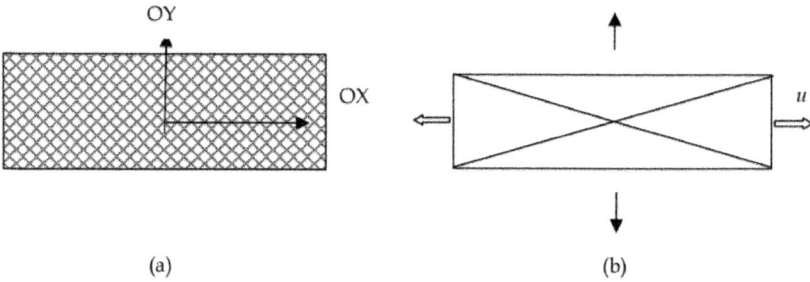

(a)   (b)

Figure 3: Structure (a) and the mesofragment (b) of obliquely reinforced auxetic composite

At $E_2 << E_1$ we get $v_{xz} \approx 1 - ctg^2\theta$, where E_1, E_2 are Young's moduli of the fibres and matrix respectively. For small fibre packing angles, θ, Poisson's ratio v_{xz} has negative values. The deformation results in a pantographic change in orientation of the fibres, which elongate insignificantly compared to the low-modulus matrix, thus promoting its contraction normal to the reinforcement direction. To this class of auxetics belong the laminates produced by the oblique superposing of the layers (Figure 3b). Investigations into laminates made of prepregs with carbon fibres and epoxy matrix have shown that vxz of the composite obtained at small packing angles of the layers (10°-40°) is negative

Analysis of Contact Deformation of Auxetic Composites

The stress state parameters were determined for the double-lap type joint in conditions of initial compression δ_y and compression with shear δ_x (Figure 4a). The analysis of the auxetic element 1 interaction with two conjugated and located symmetrically non-deformable bodies 2 and 3 (Figure 4b) has been carried out using the finite element solution of contact problem with friction.

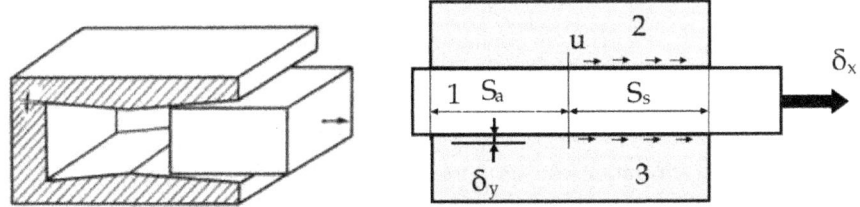

Figure 4: General view (a) and calculation scheme (b) of a frictional joint with auxetic element.

A peculiarity of this problem is a considerable nonlinear deformation brought about in conditions of unlimited shear by formation of the zones of adhesion S_a and slippage S_s with nonzero tangential contact displacements u (Figure 4b). Under compression of the joint the slippage zones are located symmetrically to the central zone of adhesion. Shear application leads to violations of this symmetry. The limiting load capacity of the joint is dependent on slippage onset over the whole contact area which, in its turn, is dependent upon the material compressibility. For the case of a quasi-isotropic material Poisson's ratio was varied within theoretically acceptable values of the isotropic elastic medium, i.e. $-1\ 0 \le\ \le v\ ,5$. The extreme values of the contact stresses tend to localise near to the right edge of the junction. The contact parameters vary insignificantly for the positive Poisson's ratios typical of isotropic materials, except for the limiting values characteristic for practically incompressible elastomers. Incompressibility ($v = 0.48-0.5$) results in the elastic compression of the material and contact slippage. The stress strain state parameters including the maximal equivalent stress σ_{eqv}, contact pressure p, tangential stress τ and slippage u, have been studied as a function of Poisson's ratio for the quasi-isotropic materials and the reinforcement angle for the anisotropic ones (Figure 5). An abrupt leap in the maximal contact parameters is observed when $v < 0$, not seen with the positive Poisson's ratios. This increase is most marked when $v < -0.9$. The adaptive mode of friction has been studied in the form of a self-locking effect under contact loading in isotropic and anisotropic auxetic cases [Shilko et al., 2008a]. This effect suggests that the strength of such a joint rises with increasing shear load.

Similar calculations have been made for a joint with a deformable element of the anisotropic auxetic composite on the low-modular matrix base (see section 3.2, Fig. 3) under varying reinforcement angles that determine the elastic moduli E_x, E_y, E_z, v_{xy}, v_{yz}, v_{xz}, G_{xy}, G_{yz}, G_{xz}. These elastic constants (Table 4) were calculated based on the volume fraction of the fibrous filler with $\mu = 0.1$, elasticity modulus and Poisson's ratio of the matrix and filler, respectively $E_m = 4$ MPa, $v_m = 0.5$, $E_f = 1.5$ GPa, $v_f = 0.4$ using the formulas

$$E_x = \frac{1}{a_{11}}; \; E_y = \frac{1}{a_{22}}; \; E_z = \frac{1}{a_{33}}; \; G_{xy} = \frac{1}{a_{66}}; \; G_{yz} = \frac{1}{a_{55}}; \; G_{xz} = \frac{1}{a_{44}};$$

$$v_{xy} = -E_x a_{12}; \; v_{yz} = -E_y a_{23}; \; v_{xz} = -E_x a_{13}. \tag{1}$$

where a_{ij} are compliance coefficients of a unidirectional composite:

$$a_{11} = \frac{1}{(1+(n-1)\mu)E_m}, \; a_{22} = a_{33} = \frac{\left(\mu+n(1-\mu)(1+(n-1)\mu)-\left(nv_m-v_f\right)^2\mu(1-\mu)\right)}{(1+(n-1)\mu)E_f},$$

$$a_{12} = a_{13} = -\frac{v_m(1-\mu)+v_f\mu}{(1+(n-1)\mu)E_m}, \; a_{23} = -\frac{\left(v_f\mu+nv_m(1-\mu)\right)(1+(n-1)\mu)+\left(nv_m-v_f\right)^2\mu(1-\mu)}{(1+(n-1)\mu)E_f},$$

$$a_{66} = a_{55} = 2\frac{(1+v_m)\left(n(1+v_m)(1-\mu)+(1+v_f)(1+\mu)\right)}{\left(n(1+v_m)(1+\mu)+(1+v_f)(1-\mu)\right)E_m}, \; a_{44} = 2\frac{(1+v_f)\mu+n(1+v_m)(1-\mu)}{E_f} \tag{2}$$

Where $n = E_f / E_m$.

The minimal Poisson's ratio $v_{xz} = -2.142$ is attained when the angle between the reinforcement direction and OY axis is 70° and deformation u is directed along OX axis (Figure 3).

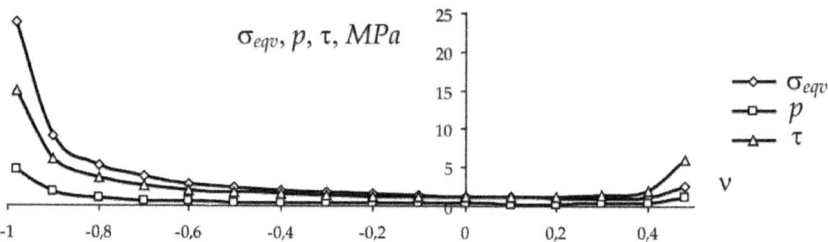

Figure 5: Maximal values of equivalent stresses, contact pressures and tangential stresses as dependent on Poisson's ratio

The maximal contact parameters were determined for different surface geometries of the conjugated bodies, namely plane, cylindrical (curvature radius r = 100 mm) and wedge-like (wedge aperture angle α = 174°). It is seen in Figs. 6-10 that the extreme dependence of contact stresses is characteristic for all geometries with a minimum at reinforcement angle 45°.

φ, degree	0	15	30	45	60	70	80	90
E_x, MPa	6.627	6.286	5.619	7.788	38.47	133.4	266.8	303.2
E_y, MPa	303.2	207.5	38.47	7.788	5.619	6.051	6.471	6.627
E_z, MPa	6.627	7.259	11.96	81.50	11.96	7.95	6.88	6.627
v_{xy}	0.010	0.082	0.336	0.945	2.242	2.793	0.912	-0.48
v_{yz}	0.480	-1.727	-1.316	0.048	0.662	0.858	0.958	0.989
v_{xz}	0.989	0.918	0.662	0.048	-1.316	-2.142	-0.743	0.48
G_{xy}, MPa	1.996	21.1	59.58	78.17	63.31	42.49	16.98	1.996
G_{yz}, MPa	1.996	2.285	3.181	3.754	2.753	2.521	1.771	1.666
G_{xz}, MPa	1.666	1.910	2.753	3.754	3.181	2.117	2.122	1.996

Table 4: Elastic constants for obliquely reinforced composite as a function of reinforcement angle φ

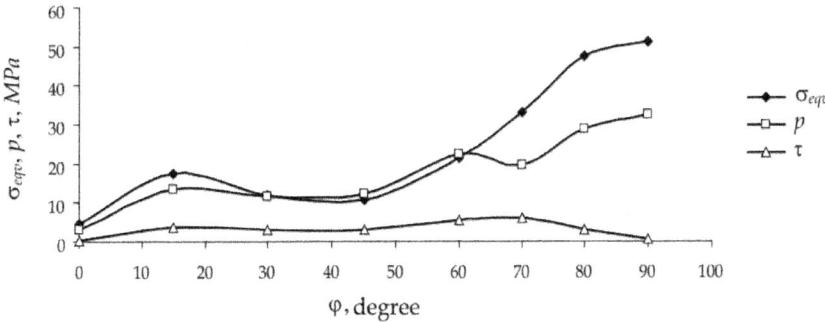

Figure 6: Dependence of maximal values of equivalent stress, contact pressure and tangential stress on reinforcement angle at compression: $\delta_y = -1$ mm (plane surface).

With a plane surface of the conjugated bodies (Figures 6, 7) the dependencies of stresses $\sigma_{eqv}(\varphi)$, $p(\varphi), \tau(\varphi)$ have local maxima. Their location varies with increasing shear due to slipping.

For cylindrical conjugated bodies, the local minima are absent under pure compression of the auxetic section, although shear promotes their appearance in the region of large reinforcement angles (Figures 8, 9).

It is peculiar that the auxetic body in a junction with a wedge surface shows a rather weak shear effect upon the contact stress state. Similarly to the case of planar surfaces, the local minima of maximal stress σeqv and contact pressure p correspond to the reinforcement angle of φ = 15°.

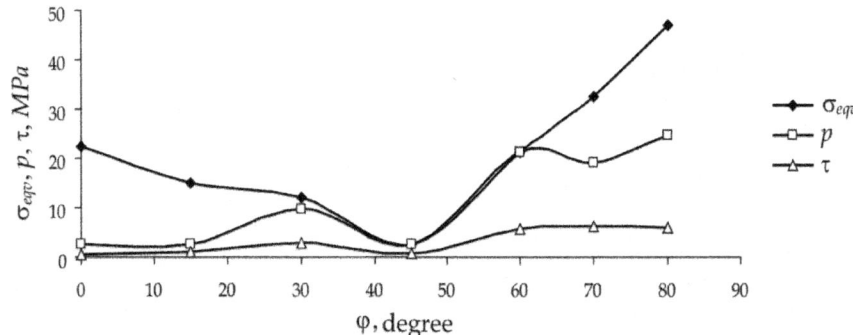

Figure 7: Dependence of maximal values of equivalent stress, contact pressure and tangential stress on reinforcement angle at compression with shear: $\delta_y = -1$ mm, $\delta_x = 5$ mm (plane surface)

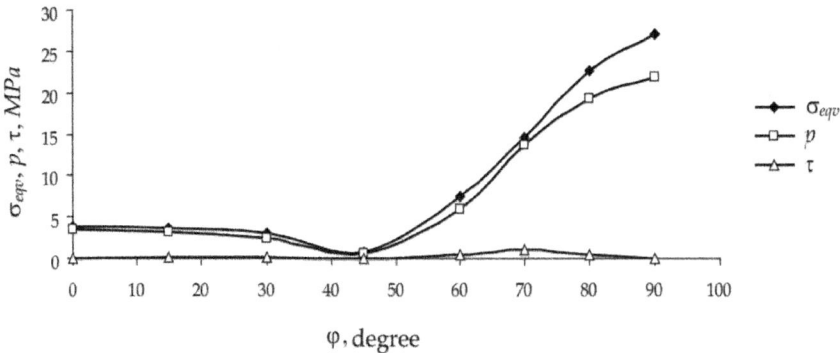

Figure 8: Dependence of maximal values of equivalent stress, contact pressure and tangential stress on reinforcement angle at compression: $\delta_y = -1$ mm (cylindrical surface)

So, promising functional materials with negative Poisson's ratios (auxetics) have been considered. The results reported here help to quantitatively evaluate the influence of Poisson's ratio (in the isotropic materials) and reinforcing angle (in the anisotropic composites) for compression and compression with shear contact interactions. The adaptive mode of friction has been studied in the form of self-reinforcing under contact loading in isotropic and anisotropic auxetics. This effect suggests that the bearing capacity of such a frictional joint rises with increasing shear [Shilko & Stolyarov, 1996]. It is shown that the use of auxetic materials is an efficient means of improving the mechanical and frictional characteristics of composites.

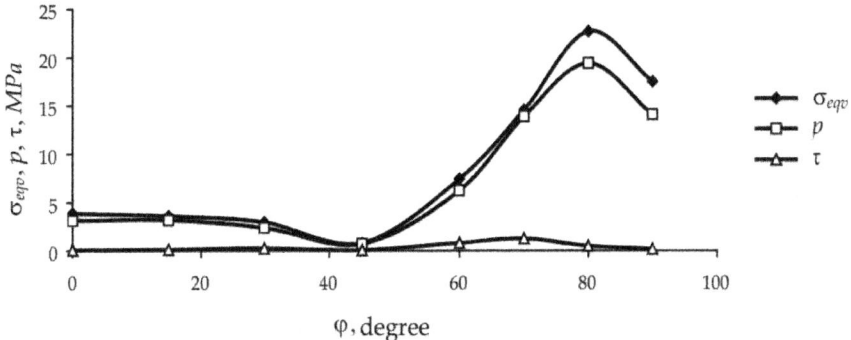

Figure 9: Dependence of maximal values of equivalent stress, contact pressure and tangential stress on reinforcement angle at compression with shear: $\delta_y = -1$ mm, $\delta_x = 0.1$ mm (cylindrical surface).

SELF-STRUCTURING OF MULTIMODULE MATERIALS

As it was mentioned above, in contrast to traditional composites, which display stable interfaces between components determining invariability of technologically specified complex of properties (Table 1), the adaptive material admits mobility of interfaces. As a consequence, description of able to regulate its structure ACM proceeding from the given optimal criterion presumes statement of the moving boundaries problem and use of methods for its solution. In this connection, let's turn to investigation results of elastic (reversible) remodelling of a physically nonlinear multimodular metals such as Fe, Al, Cu, Mg, etc [Bell, 1968] (particularly, in Figure 10 these data are given for Fe). Softening of bulk modulus near the volume phase transition has been observed in polymer gels too [Hirotsu, 1991]. In the paper [Baughman & Galvao, 1993] it has been shown that crystalline networks demonstrates unusual mechanical and thermal properties. The hypothesis is to be introduced into the course of the model development according to which the adaptive reaction reduces is a specific transfer process to an optimum control over intrinsic to ACM moving boundaries.

Let a composite at the polycomponental level be formed by bonded elastic particles (Fig. 11a). Being initially homogeneous and isotropic, each particle is characterized under the force action by stress concentration in the contact zone with the neighboring particle (Fig. 11b). The role of the sensor function is presumed to be played by characteristic for a number of substances multimodule ability [Bell, 1968], that is, the presence of a set of n discrete values of Young's modulus E depending on stress σ, being in this case the control parameter.

For numeric modeling of adaptation to extreme external loads, which are by far higher the acceptable one for the initial material, discretization of each particle by finite elements and block relaxation algorithm, are used. A system of inequalities is taken as a processor function of the multimodule composite, where the lower and upper estimates of stresses are corrected at model "exposure" proceeding from the optimum criterion, namely, the condition of the minimal equivalent stress σ_{eq}

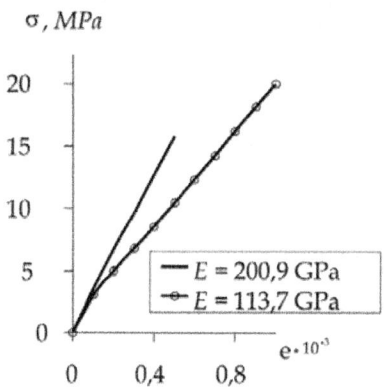

Figure 10: Stress–strain curve of multimodule material (Fe) having two values of elastic modulus E.

$$E = \begin{cases} E_1, & \text{if } \sigma^0 \leq \sigma \leq \sigma^1; \\ \text{-----------} \\ E_l, & \text{if } \sigma^{l-1} \leq \sigma \leq \sigma^l; \\ \text{-----------} \\ E_n, & \text{if } \sigma^{n-1} \leq \sigma \leq \sigma^n, \end{cases}$$

$$(3)$$

Simulation results prove that the adaptive reaction consists in transformation of the initial homogeneous structure into a reversible inhomogeneous one (Fig. 11c,d) ensuring perception of extreme loads through effective reduction of stress concentration due to dynamically optimal distribution of elastic modulus E. So, the statement and systematic analysis of the problem of developing adaptive composites has enabled to trace evolution of structural organization of artificial materials, to clarify mechanisms of adaptation to external media and to disclose, to a certain degree, the effect of structure on formation of the optimum back reaction. In above considered example simulation of composite materials adaptivity is formulated as a problem on localizing moving internal boundaries, while differentiation of material functions is related to the changing scale level of the structure.

SELF-ASSEMBLING OF AUXETIC POROUS COMPOSITES WITH MULTIMODULAR SOLID PHASE

Porous or cellular materials like «solid – gas» inhomogeneous systems are efficient composite structures in respect to optimizing strength and stiffness for a given weight.

These materials are useful for cushioning, insulating, damping, absorbing the kinetic energy from impact, packing, etc. Stiff and strong ones are preferable in load-bearing structures such as a lightweight core in sandwich panels. The term cellular is appropriate when the material contains polyhedral closed cells, as if it had resulted from solidification of a liquid foam.

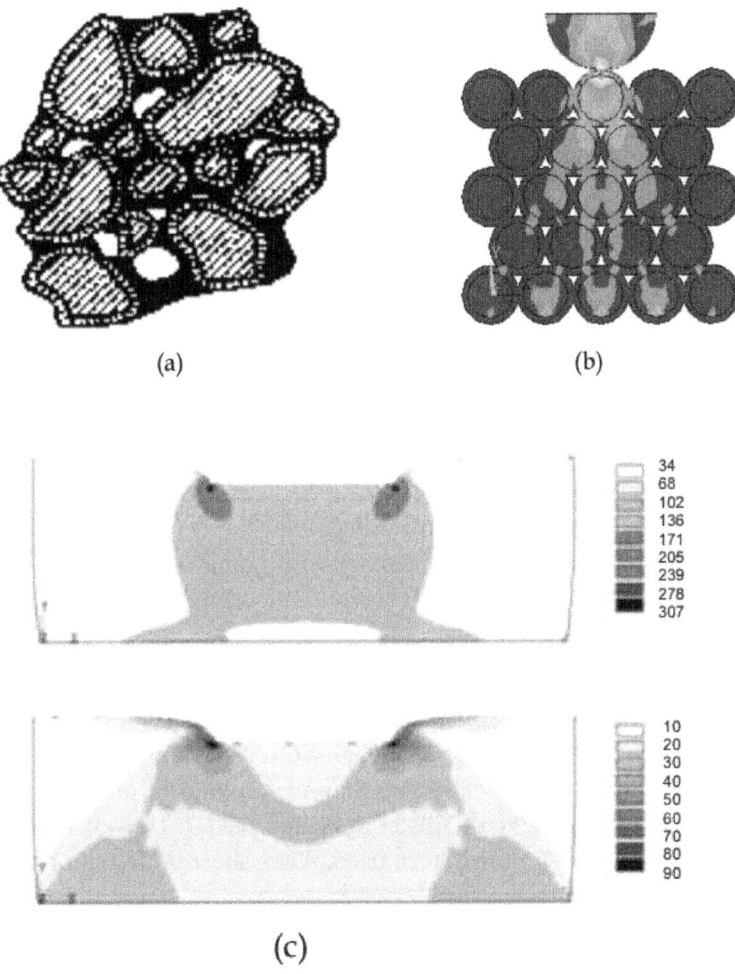

(a)   (b)

(c)

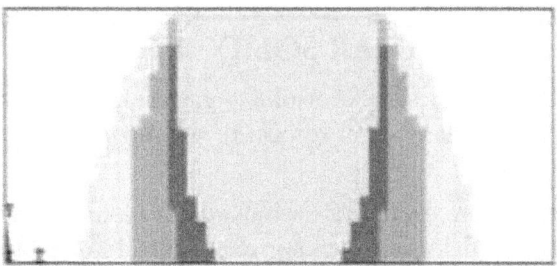

Figure 11: ACM response after loading: typical composite structure (a); mesomechanical model (b); initial and final (after remodeling) distribution of equivalent stresses σ_{eq}. (c); dynamically optimal distribution of elastic modulus E (d)

A new means of improving the mechanical characteristics may be realized using abnormal deformation properties of auxetic porous materials having a negative Poisson's ratio ν [Lakes, 1987]. Auxetic porous materials, including auxetic porous nanomaterials (APN), having very high mechanical properties, are suitable for creating adaptive contact joints and for replacing natural materials such as damaged bone and tooth biotissues.

Examination of the mechanisms of generating a negative Poisson's ratio has been discussed and published in the last years, including special issues of physica status solidi (b) journal [Wojciechowski et al, 2007]. It is known that the inverted or re-entrant cell structure of porous auxetics may be obtained by isotropic permanent volumetric compression of the conventional foam, resulting in non-reversible micro-buckling of the cell walls. There is interest in compression-driven self-assembly as a means to create auxetic porous structures at the nanoscale. Below we predict the deformation behaviour of porous materials under uniaxial tension and compression (by an analytical method) and contact compression (by the finite element method).

Analytical Determination of Porous Material Elastic Modulus

For open-cell flexible cellular materials, Poisson's ratio can be determined by a rod type structural unit with chaotically oriented cubic cells, as presented in Figure 12. It is worth mentioning that such a kind of unit cell model has been simulated in reference [Gibson & Ashby, 1982]. However, a cubic, not a spherical, structural unit had been used. Also, shear deformation of the rods had not been taken into account.

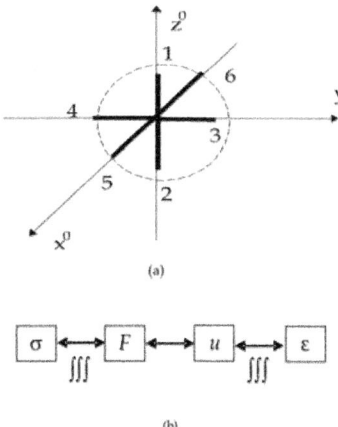

(a)

(b)

Figure 12: Structural unit (a) and simulation procedure (b) of flexible cellular plastics: σ, ε are stress and strain tensor components; F is force acting on the rod end; u is displacement of force application point relatively to the rods; $\iiint$ is averaged over direction.

The rods of this structural unit are directed normally to the cubic planes. Symmetry of the element allows one to represent the displacement of the force application points (ends of rods) relatively to the rod joints through the deformation tensor components.

$$\Delta x_{L1} = x_{L1} - L = L\varepsilon_{z^0 z^0}, \quad y_{L1} = \frac{L}{2}\sqrt{\gamma_{x^0 z^0}^2 + \gamma_{y^0 z^0}^2},$$

$$\Delta x_{L3} = x_{L3} - L = L\varepsilon_{y^0 y^0}, \quad y_{L3} = \frac{L}{2}\sqrt{\gamma_{x^0 y^0}^2 + \gamma_{y^0 z^0}^2},$$

$$\Delta x_{L5} = x_{L5} - L = L\varepsilon_{x^0 x^0}, \quad y_{L5} = \frac{L}{2}\sqrt{\gamma_{x^0 y^0}^2 + \gamma_{x^0 z^0}^2}, \tag{4}$$

where L is the structural unit rod length; x_{Li}, y_{Li} are the coordinates for the end of the i-th rod (i = 1..6) in the xy coordinate system; the x axis is directed longitudinally to the i-th rod in the non-deformed state (Figure 13).

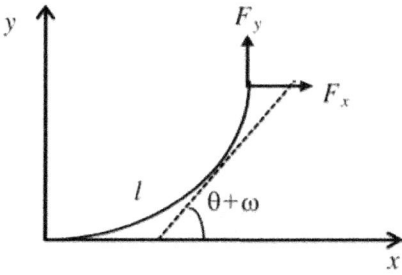

Figure 13: Scheme of cantilever beam under large bending.

Eq. (4) refers to deformations for which the Cauchy relations are satisfied. Here, the parameter L can be related to the solid state volumetric fraction by the following equation

$$V_f = \frac{V_m}{V} = \frac{9}{2\pi q^2},$$

(5)

where V_m is the volume of rods in the structural unit; V is the structural unit total volume before deformation; q is the rod length L to its cross sectional side length r ratio. For simplification we neglected the volume of the nodes (rod joints) and assumed that the rods have a square cross-section. During further calculations we have estimated that the simulation results do not depend on the r value. So we may assume that r = 1, L = q.

Let us assume that in the coordinate system XYZ uniaxial strain is defined as ( ) ε nm = f t (other components of strain are equal to zero). The system XYZ position relative to the system $x^0 y^0 z^0$ is defined by Euler's angles β^1, β^2, β^3. Once the function f(t) and Euler's angles are known, these define the deformation components in the $x^0 y^0 z^0$ system (Fig. 12). Then, the displacements (4) can be written as follows:

$$y_{Li} = \varepsilon_{nm}(t)\eta_i(\beta_1,\beta_2\beta_3), \quad \Delta x_{Li} = \varepsilon_{nm}(t)\xi_i(\beta_1,\beta_2\beta_3).$$

(6)

Here $\eta_i(\beta_1, \beta_2, \beta_3)$, $\xi_i(\beta_1, \beta_2, \beta_3)$ are the Euler's angle functions which are related by recalculating the tensor components under coordinate axis rotation. For the determination of forces Fi f acting at the ends of the rods by the set deflections, it is necessary to solve a large flexure problem of a cantilever beam taking into account material viscosity. At the same time, to describe deformation of the low-density porous materials ($V_f < 0.1$) it can be assumed that the rod is deformed equally over all length L. The viscoelastic behaviour of the rod material is described by Rzhanitsyn's relaxation function

$$R(t) = Ae^{-\beta t}t^{\alpha-1},$$

(7)

where t is time, s; and A, α, β are the kernel parameters. The stress/strain relations are determined by the following equation

$$s_{\rho\chi} = 2G_f\left(\upsilon_{\rho\chi} - \int_0^t R(t-\tau)\upsilon_{\rho\chi}(\tau)d\tau\right), \quad \sigma = 3K_f\varepsilon.$$

(8)

where $s_{\rho\chi}$, $\upsilon_{\rho\chi}$, σ, ε are the deviatoric and spherical parts of the stress and strain tensors; Gf, Kf are the shear and bulk moduli of the material. For the beam deformations, let us assume

$$\varepsilon_{ll} = \varepsilon_0(l) + \lambda\theta'(l), \quad \varepsilon_{\lambda l} = \frac{1}{2}\omega(l).$$

$$(9)$$

where l is the coordinate referred along the rod median in the deformed state; λ is the coordinate referred perpendicularly to l; θ is the rotation of the rod cross-section connected with flexural strain; θ '– θ is the derivative of the l coordinate; ω is the rod cross-section turning angle as a function of shear strain; ε_0 is the deformation of the centre line passing through the centre of gravity under tension or compression.

The allowance for flexural, shear and tensile-compression strains helps to describe deformation of "short" rods when their length is commensurable with the cross-sectional side length. For an arbitrary cross-sectional shape, the following expressions are valid

$$M = \iint_S \sigma_{ll}\lambda dS, \quad P = \iint_S \sigma_{ll}dS, \quad Q = \iint_S \sigma_{\lambda l}dS .$$

$$(10)$$

where M is the bending moment; Q, P are the transverse and longitudinal forces. Therefore, the equilibrium Eqs. for the cantilever rod for the large deflection case will take the form

$$Q = F_y \cos(\theta + \omega) - F_x \sin(\theta + \omega),$$
$$P = F_x \cos(\theta + \omega) + F_y \sin(\theta + \omega),$$
$$M = F_y(x_L - x) - F_x(y_L - y).$$

$$(11)$$

Substituting Eqs. (8) and (9) into (10) gives

$$\omega = \frac{-k}{G_f S}(F_x \sin(\theta + \omega) - F_y \cos(\theta + \omega)) + \int_0^t R(t - \tau)\omega(\tau)d\tau,$$

$$\varepsilon_0 = \frac{1}{E_f S}(F_y \sin(\theta + \omega) + F_x \cos(\theta + \omega)) + \int_0^t R(t - \tau)\varepsilon_0(\tau)d\tau,$$

$$\theta' = \frac{1}{E_f J}\left(F_y(L + \varepsilon_{nm}\xi - x) - F_x(\varepsilon_{nm}\eta - y)\right) + \int_0^t R(t - \tau)\theta'(\tau)d\tau,$$

$$x' = \cos(\theta + \omega),$$

$$y' = \sin(\theta + \omega).$$

$$(12)$$

where J, S are the second moments of the area and the cross-sectional area of the rod, correspondingly; Ef is Young's modulus of rod material; k is the coefficient complying with non-uniformity of tangential stress distribution over the cross-sectional area. In our calculations we assumed k = 1.

Therefore, a system of Eqs. was obtained for the five unknown coordinates l and time functions. Let us apply the following boundary conditions: $\theta(0,)$ $(0,)$ $(0,)$ 0 t xt yt $= = =$. In (12) η, ξ are constants. Solution of these combined Eqs. using the finite difference method allows us to obtain the coordinates of the free end of rod as a function of five variables, viz:

$$x_L = x(L,t) = f_x(F_x, F_y, \eta, \xi, t),$$
$$y_L = y(L,t) = f_y(F_x, F_y, \eta, \xi, t).$$

$$(13)$$

During computation of (12) it was taken into account that the l coordinate differentiation is made in the deformed state. Therefore, the increment of the l parameter was assumed equal to $dl = (1+\varepsilon_0)\dfrac{L}{n_0}$. Here, n_0 is a discretization number. The solution of (12) was carried out for a specified t. It should be mentioned that the structure of Rzhanitsyn's relaxation function (7) causes the integral terms in (12) to contain θ, γ and ε_0 functions which were defined during the previous steps

The conditions for calculation of the required forces are of the type

$$\begin{cases} f_x(F_x, F_y, \eta, \xi, t) = L + \varepsilon_{nm}(t)\xi, \\ f_y(F_x, F_y, \eta, \xi, t) = \varepsilon_{nm}(t)\eta. \end{cases}$$

$$(14)$$

The solution of Eqs. (12) and (14) was obtained numerically with the help of MathCad® 7.0 software. The system of nonlinear Eqs. was solved using Newton's method. As the initial approximation we took the solution of the previous step. Therefore, we obtain the functions $F_x(\eta,\xi,t)$, $F_y(\eta,\xi,t)$ which can be presented as follows

$$F_x = C_{x1}\xi + C_{x2}\eta + C_{x3}\xi\eta + C_{x4}\xi^2 + C_{x5}\eta^2 +$$
$$+C_{x6}\xi^2\eta + C_{x7}\xi\eta^2 + C_{x8}\xi^3 + C_{x9}\eta^3 + C_{x10}\xi^2\eta^2.$$

$$(15)$$

At a given t, the coefficients C_{xj}, C_{yj} ($j = 1..10$) can be defined by standard regression procedures. The stress tensor components are related through the forces (15) as follows:

$$\sigma_{x^0x^0} = F_{x1}\frac{1}{\pi(L + \varepsilon_{nm}\xi_2)(L + \varepsilon_{nm}\xi_3)},$$
$$\sigma_{y^0y^0} = F_{x2}\frac{1}{\pi(L + \varepsilon_{nm}\xi_1)(L + \varepsilon_{nm}\xi_3)},$$

$$(16)$$

$$\sigma_{z^0z^0} = F_{x3}\frac{1}{\pi(L+\varepsilon_{nm}\xi_1)(L+\varepsilon_{nm}\xi_2)},$$

$$\sigma_{x^0y^0} = F_{y1}\frac{1}{\pi(L+\varepsilon_{nm}\xi_2)(L+\varepsilon_{nm}\xi_3)}\frac{\varepsilon_{x^0y^0}}{\left(\varepsilon_{x^0y^0}^2+\varepsilon_{x^0z^0}^2\right)^{1/2}},$$

$$\sigma_{x^0z^0} = F_{y1}\frac{1}{\pi(L+\varepsilon_{nm}\xi_2)(L+\varepsilon_{nm}\xi_3)}\frac{\varepsilon_{x^0z^0}}{\left(\varepsilon_{x^0y^0}^2+\varepsilon_{x^0z^0}^2\right)^{1/2}},$$

$$\sigma_{y^0z^0} = F_{y2}\frac{1}{\pi(L+\varepsilon_{nm}\xi_1)(L+\varepsilon_{nm}\xi_3)}\frac{\varepsilon_{y^0z^0}}{\left(\varepsilon_{x^0y^0}^2+\varepsilon_{y^0z^0}^2\right)^{1/2}}.$$

$$(17)$$

The stresses for the XYZ system were then redefined. Because of the chaotic orientation of the unit cells, the stress tensor components should be averaged over direction (Euler's angles)

$$\sigma_{nm} = \int_0^\pi\int_0^{2\pi}\int_0^{2\pi}\sigma_{nm}(\beta_1,\beta_2,\beta_3)\frac{\sin\beta_3}{8\pi^2}d\beta_1 d\beta_2 d\beta_3.$$

$$(18)$$

Therefore, for the known stress to time dependence, we defined time dependencies of the stresses in a representative volume of the material.

Calculation Example: Uniaxial Stress

As an example of using above technique, let us examine the stress-strain state of an elastic porous material based on high density polyethylene (HDPE). Experimental data for HDPE were obtained from [Goldman, 1979]: $G_f = 237$ MPa; $K_f = 1402$ MPa; $A = 0.022$ s- β ; $\beta = 2.995 \cdot 10\text{-}5$ s-1; $\alpha = 0.175$.

Averaging in all possible loading directions (14) makes the simulated material isotropic at the macroscopic level. The $\tau(\gamma)$ function therefore characterizes the dependence of stress on strain deviator components $\tau(\gamma) = s_{nm}(2\upsilon_{nm})$. Thus, if functions $\tau(\gamma)$ and $p(\Theta)$ are known, it is possible to simulate isotropic material behaviour at an arbitrary homogeneous stress-strain state. Hence, for a uniaxial stress ($\sigma_{ZZ} \neq$) the following relations are true

$$2\upsilon_{ZZ} = \frac{4}{3}\varepsilon_{ZZ}(1+\mu);$$

$$\Theta = (1+\varepsilon_{ZZ})(1-\varepsilon_{ZZ}\mu)^2 - 1;$$

$$s_{ZZ} = \frac{2}{3}\sigma_{ZZ};$$

$$\sigma = \frac{1}{3}\sigma_{ZZ}.$$

$$(19)$$

We introduce the transverse strain factor $\mu = -\frac{\varepsilon_{XX}}{\varepsilon_{zz}}$, which is analogous to Poisson's ratio in the linear elasticity region. Making allowance for a large bending flexure of the ribs where μ depends on strain ε_{ZZ}, this dependence is determined by the following Eq.

$$\tau\left(\frac{4}{3}\varepsilon_{ZZ}(1+\mu)\right) = 2p\left((1+\varepsilon_{ZZ})(1-\varepsilon_{ZZ}\mu)^2 - 1\right).$$

(20)

Under stretching, μ also decreases rapidly when the strain reaches εcr. In addition, the $\mu(\varepsilon_{ZZ})$ dependence rapidly passes on the horizontal plateau $\mu(\varepsilon_{ZZ}) = const = v^0$, where Poisson's ratio v^0 is defined as

$$v^0 = \frac{3K - 2G^0}{6K + 2G^0},$$

here K is the foam bulk modulus defined by the initial part of the $p(\Theta)$ curve; G0 is the shear modulus defined by the $\tau(\gamma)$ curve.

It was found that the $\mu(\varepsilon_{ZZ})$ function does not depend on the strain rate. In Figure 14 the dependence of the factor μ on the longitudinal strain ε_{ZZ} at stretching (a) and compression (b) of an elastic cellular plastic based on HDPE ($V_f = 0.01$) is presented. Under compression the strain reaches some critical value εcr and μ rapidly decreases and becomes negative at $\varepsilon_{ZZ} > 0.9\%$. Such an anomaly of the elastic behaviour was experimentally observed in polymer foams [Lakes, 1987]. Our investigations showed that this effect may occur in cellular materials with a tetrahedronal cell form when the cell ribs buckle inward or in a honeycomb microstructure.

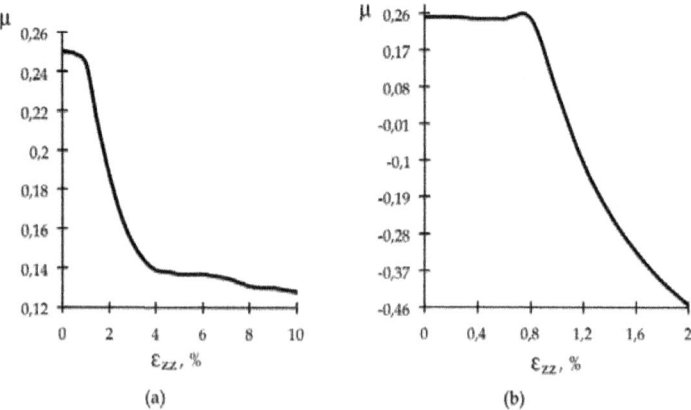

(a)   (b)

Figure 14: Dependence of transverse strain factor μ on longitudinal strain εZZ under stretching (a) and compression (b) of flexible cellular plastics.

At small strains, μ remains constant and coincides with Poisson's ratio. Under compression the strain reaches some critical value ε_{cr} and μ rapidly decreases and becomes negative at ε_{zz} > 0.9%. Under stretching, μ decreases rapidly when the strain reaches εcr and $\mu(\varepsilon_{zz})$ dependence passes on the horizontal plateau.

By defining the $\mu(\varepsilon_{zz})$ function, the dependence of stress σ_{zz} on strain ε_{zz} can be obtained as

$$\sigma_{ZZ}(\varepsilon_{ZZ}) = \frac{3}{2}\tau\left(\frac{4}{3}\varepsilon_{ZZ}\left[1 + \mu(\varepsilon_{ZZ})\right]\right).$$

(22)

At small strains, stability of μ allows us to determine the correlation between $\varepsilon_{cr}, \Theta_{cr}$ and γ_{cr}

$$\varepsilon_{cr} = \frac{1}{1-2v}\Theta_{cr}, \quad \gamma_{cr} = \frac{4(1+v)}{3(1-2v)}\Theta_{cr}.$$

(23)

Comparison with Experimental Data

To examine the applicability of the theoretical model for foam deformation properties, we compared the calculated and experimental values of the relative Young's modulus E/E_f and critical strains ε_{cr} proceeding from the following considerations: the majority of experimental data on deformation of elastic foams are based on their uniaxial compression behaviour; the calculated stress/strain dependence and the experimental behaviour are almost linear at $\varepsilon < \varepsilon_{cr}$. As it was shown in [Hilyard & Cunningham, 1987], $\sigma_{zz}(\varepsilon_{zz})$ dependence at ε_{zz} > ε_{cr} to a certain degree is conditioned by inhomogeneity of the inner structure of the material.

During definition of Young's modulus E of an elastic cellular plastic, we considered that the rod cross-section turning angles are small $(\cos(\theta + \omega) \approx 1, \sin(\theta + \omega) \approx 0)$ and we do not consider rod viscosity. In this case, the solutions can be obtained in the analytical form. For the relative Young's modulus, we have

$$\frac{E}{E_f} = V_f \frac{36 + V_f\pi(7 + 4v_f)}{216 + 3V_f\pi(9 + 8v_f)},$$

(24)

where v_f is the solid phase Poisson's ratio. In particular, for the elastic polymer material we assume that $v_f = 0.49$. Equations (10) and (15) yield an approximate expression for the critical strain ε_{cr}

$$\varepsilon_{cr} = \frac{V_f \pi^3 \left[72 + V_f \pi \left(9 + 8 v_f \right) \right]}{72 \left[36 + V_f \pi \left(7 + 4 v_f \right) \right]} \cdot$$

(25)

The dependence of the relative Young's modulus E/E_f on the relative solid volume fraction V_f for the elastic foam is presented in Figure 15.

In Figure 15, curve 1 corresponds to Eqs. (23). Curve 2 agrees with the results obtained in [Warren & Kraynik, 1987]. Curve 3 meets the results obtained in [Beverte & Kregers, 1987] using the semi-axes hypothesis. Curve 4 corresponds to the analytical expression

$$\frac{E}{E_f} = \frac{V_f}{3} \left(1 - 2 v^{/} \right) = 0,16 V_f ,$$

(26)

obtained in [Gibson & Ashby, 1982]. Here $v /$ is the Poisson's ratio of the material dependent on the number of rods in structural unit N. For simulation of mechanical behaviour of the rubber foam [Gibson & Ashby, 1982], we used a structural element with $4 < N < 8$, when $v / = 0.26$. Curve 5 in Figure 15 corresponds to the empirical relation for the relative Young's modulus of foam rubbers [Hilyard & Cunningham, 1987]

$$\frac{E}{E_f} = \frac{V_f}{12} \left(2 + 7 V_f + 3 V_f^2 \right) .$$

(27)

The circles in Figure 15 reflect experimental data for the foam rubber [Lederman, 1971]. This comparison proves that the proposed technique makes it possible to predict quite accurately elastic properties of the material at $V_f < 0,15$.

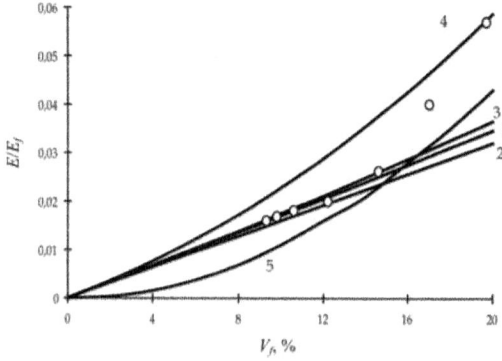

Figure 15: The dependence of relative Young modulus E/Ef on the solid phase volumetric fraction Vf for the flexible foam: curve 1 corresponds to Eqs. (23); curve 2

corresponds to results obtained in [Warren & Kraynik, 1987]; curve 3 corresponds to results obtained in [Beverte & Kregers, 1987]; curve 4 corresponds to results obtained in [Gibson & Ashby, 1982]; curve 5 corresponds to results obtained in [Hilyard & Cunningham, 1987]; circles correspond to experimental data [Lederman, 1971]

Construction of The Mesomechanical Model

Mesomechanical (in the scale of the separate cells) description of cellular structure is timeconsuming but a very informative method. A possibility for the determination of Poisson's ratio during special thermomechanical treatment of basic porous material when convex structural cells transform into concave ones as shown in Figure 16, is an advantage of the mesoscopic model.

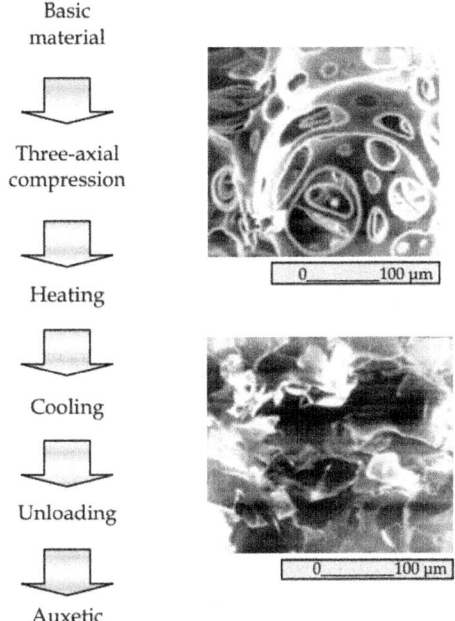

Figure 16: A scheme of obtaining the auxetic material using transformation of ther basic structure into the inverted one with concave cells. Electron microscopy of a porous polyurethane fragment with magnification 50*

Previously, the determination of v as a function of the transverse and longitudinal strain was achieved for the case of compression of the sample made of a one-phase material with known values of Poisson's ratio (Figure 1a). The geometrical sizes of the rectangular sample are $L_x = 50$ µm, $L_y = 250$ µm; the compressive strain is $\varepsilon_y = 0.5\%$. The calculated results are shown in Table 5. It should be noted that the technique has an acceptable accuracy which increases as the friction between the sample and the plates decreases. This fact is explained by a free slip of the contact surfaces.

For calculation of the effective elastic characteristics of the porous material mesofragments we replace the real structure by a system of cells of regular polyhedrons. The transformation of the porous material into the auxetic one appears to be possible at bulk compression V_{in}/V_{tr} equal to 1.4÷4.8 where V_{in}, V_{tr} are the volume of the initial and transformed structure respectively. The best results are achieved at $V_{in}/V_{tr} = 3.3÷3.7$. This agrees with the data derived for foamed polyurethane and copper sponge

The simulation allows us to describe cell transformation at the expense of free volume due to connection of structural units providing the required deformation mode.

Table 5: The calculation results of the transverse displacements

The number of node	u_x, μm			
	$f = 0.1$		$f = 0.5$	
	Left side	Right side	Left side	Right side
1	-0.0507	0.0498	-0.0510	0.0493
2	-0.0507	0.0498	-0.0510	0.0493
3	-0.0507	0.0498	-0.0510	0.0494
4	-0.0507	0.0498	-0.0509	0.0494
5	-0.0506	0.0499	-0.0508	0.0495
6	-0.0506	0.0499	-0.0508	0.0496
7	-0.0506	0.0500	-0.0507	0.0498
8	-0.0506	0.0503	-0.0508	0.0503
9	-0.0504	0.0483	-0.0504	0.0499
10	-0.0373	0.0370	-0.0471	0.0405
u_x average	-0.04929	0.04846	-0.05045	0.0487
u_x total average	0.048875		0.049575	

Mesomechanical Analysis

According to the mesomechanical approach, some systems of regular polyhedrons, presented in Figure 17, were constructed for calculation of Poisson's ratio v during structural transformation under compression (Figure 16). In the numerical example, we give the following initial data for the solid phase of the porous material E = 1 GPa, v = 0.1; the sizes of the fragment 240*280 μm and the periodic cell 34*34 μm, the friction coefficient on the contacting surface with the rigid plates is f = 0.5.

Besides the linear elastic solid phase, we have assumed a physically nonlinear multimodular solid phase. In the last case, the stepwise dependence of Young's modulus on the stress component has been used (1).

We then simulated deformation of the initial structure with rectangular cells to analyse the formation of auxetic properties under compression of traditional porous material. To increase the accuracy, Poisson's ratio was determined by averaging the displacements for the left and right sides of the model structure fragment. According to Table 6, the results in the case of a multimodule solid phase seems to be more stable than for E = const and at less expressed auxetic properties (stability loss of the porous fragment made of multimodular material is absent at compression displacement u_y = 14.0 µm).

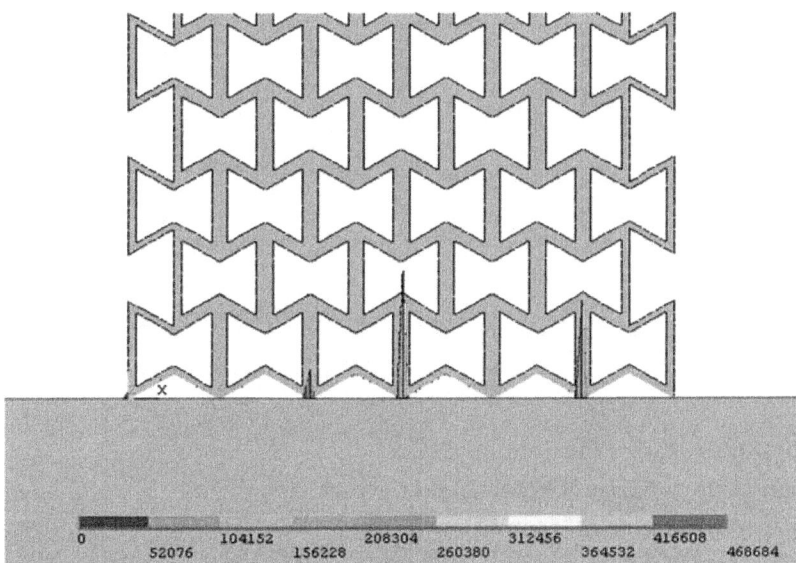

Figure 17: Distribution of contact pressure under deformation of the porous structure (the vertical compressive displacement uy = 0.1 µm).

Table 6: The calculation results of Poisson's ratio v

u_y, µm	1.4	2.8	7.0	14.0*	21.0	28.0	35.0	42.0
E = const	-0.040	-0.054	-0.085	-0.49	-0.180	-0.222	-0.130	-0.291
E = E(σ)	-0.0146	-0.0195	-0.0340	-0.076	-0.080	-0.100	-0.118	-

The dependences in Figure 18 were shown in a dimensionless form (compression level was taken as a ratio of normal displacements to the height of the porous material fragment u_y/b) for comparison under different conditions of loading. It can be seen from Figure 18 that instability of solution is observed at a step-by-step loading of the porous material with hexagonal cells at a deformation level 5%. The porous material with hexagonal cells and multimodule solid phase coincide closely. The solution is not converged for the porous material with square cells at deformation level more than 15% with local unstability in the range 2.5-7.5% and for multimodule solid phase at deformation level more than 10%. The solution is not converged at the deformation level more than 5% for the concave cells with linear elastic and multimodular solid phase. According to the previous stress history the solution is not converged at the deformation levels greater than 7.5% and 3% for the square and concave cells respectively.

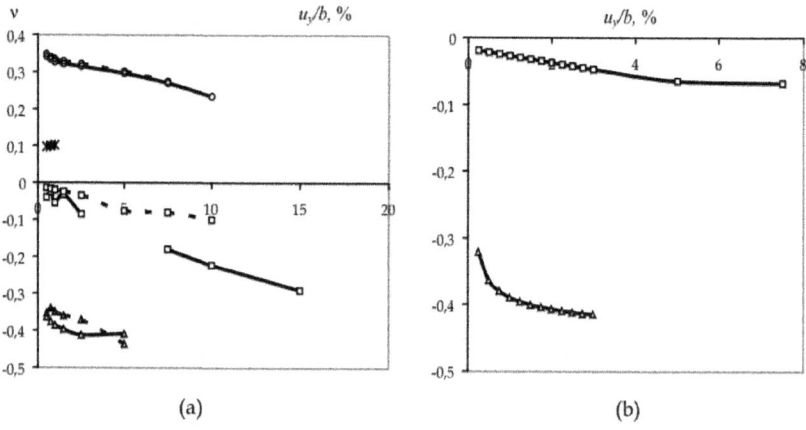

(a)   (b)

Figure 18: Dependence of Poisson's ratio on compression level: (a) step-by-step loading; (b) accounting the previous stress history for the porous material with square cells (squares), hexagonal cells (generated angle $\alpha = 600$) (triangles), circular cells (stars) and with multimodule solid phase (hatches).

The analysis of the stress-strain state of these cellular structures for various deformation levels shows that Poisson's ratio is near to zero at the initial stress state but decreases significantly under compression of the material, which its solid phase has a constant elasticity modulus. The predicted auxetic behaviour is due to generation of the concave cells at the determinative compression level. Poisson's ratio decreases for the structure with the given concave cells transferring into a plateau. At significant deformation, the solution is not converged due to closing of the cell edges. At the macroscale the model of the cell structure is unstable. This may result in the a displacement of the fragment (in the given example this takes place at compression level $u_y = 14$ μm). For obtaining a stable solution, it is necessary to take into consideration the previous stress history of the contact friction process (Fig. 18b). The account of the previous stress history is also important for calculating the auxetic self-lock mode at the conditions of contact compression and shear [Shilko et al, 2008a].

Prediction of Auxetic Effects in Porous Materials With Nano-Sized Cells

Self-assembling high-strength and rigid materials of small density are of great interest like Langmuir films. This may be reached by the auxetic porous material "construction" on the micro- and nano-size level. It is important that the value of adhesion forces F increases essentially at decreasing of the gap H between solid surfaces. The values of the adhesion force are shown in Table 3 for two pairs of polymers and three values of the gap H according to

$$F = \frac{A_{12}}{6\pi H^3},$$
(28)

where A12 is the Hamaher constant and H is the distance between surfaces. It is seen that a sharp increase of the adhesion force takes place in nano-sized cells of the porous material.

Table 3: Estimation of adhesion forces in nano-sized cells on the basis of polymers

Material	F, MPa			A_{12}, Erg
	$H = 10A$	$H = 5A$	$H = 4A$	
Polytetrafluorethylene - Polyimide	7.38	51.1	115.3	$1.39{*}10^{-12}$
Polycaproamide - Polycaproamide	7.30	58.3	113.9	$1.37{*}10^{-12}$

The calculations of the deformed state of the porous material subject to adhesion forces and multimodule effect simultaneously, show a possibility of self-assembling of a spontaneous, energetically preferable auxetic nano-sized structure as shown in Figure 19. So, the auxetic porous materials with micro- and nano-sized cells, having good combination of density, deformational and strength properties, seem quite preferable for many technical and biomedical applications. Analytical and numerical modelling describes the cellular solid transformation resulting in microbuckling of the cell walls under certain loading conditions and providing the auxetic deformation mode. Geometrically simple mesomechanical models of the porous material based on cubic, rectangular and concave structural units have been investigated in the present paper taking into account such important factors as large strains, history of loading, physical nonlinearities of solid phase, adhesive interaction and so on.

The limitations of the effective finite element simulation are caused by stability loss of the representative fragment of structure. The possibility of compression-driven self-assembly of nano-sized auxetics due to the increasing adhesion force between the cell walls has been predicted.

CONCLUSION

- The systematic analysis of the problem of developing adaptive composites has enabled us to trace evolution of structural organization of artificial materials, to clarify the mechanisms of adaptation to the external action, and to disclose, to a certain degree, the effect of structure on formation of the optimum back reaction.

- In above considered examples of composites, description of adaptive structures is formulated as a problem on localizing moving interfaces. The study of synergetic phenomena in the nonliving nature and analogous processes in biological objects will, in our opinion, provide a possibility to find structural-and-functional prototypes of adaptive composites.

- The proposed analytical and numerical models predict self-reinforcing in composites and joints made of auxetics under loading. The role of friction, previous stress history, multimodule solid phase and adhesion forces acting between the walls of cells were shown in the formation of auxetic properties of the porous materials as the composites with the gas phase.

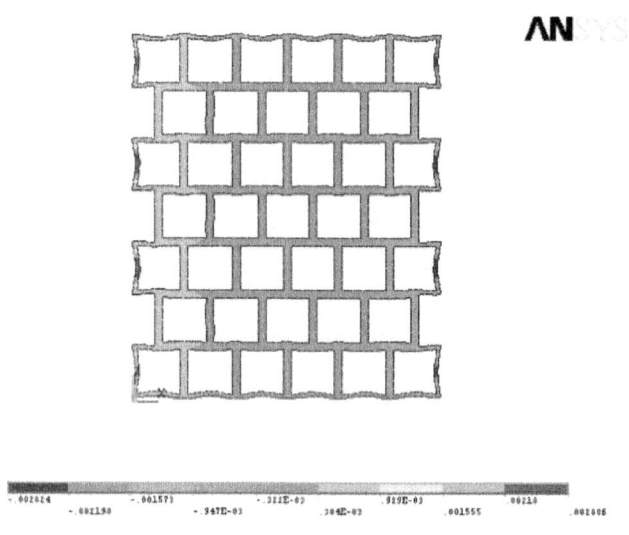

(a)

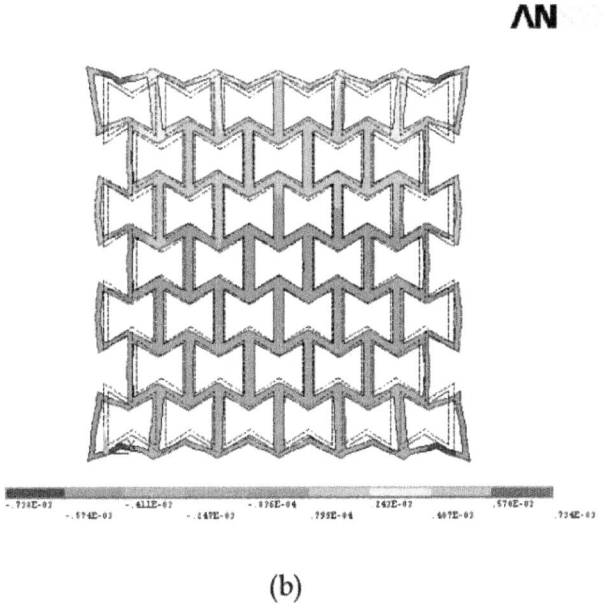

(b)

Figure 19: Deformation modes of the porous material with initially rectangular (a) and concave (b) shape of the cells under the action of adhesion forces.

- The limiting values of compression deformation on the stability criterion of cellular structures under compression and the possibility of energetically preferable selfassembly of auxetic porous nano-sized materials have been predicted.

- It's seemed that realization of self-healing in composites made of auxetic and multimodule materials is a perspective goal of further studies.

ACKNOWLEDGEMENT

The author is grateful for assistance to Prof. Yu. Pleskachevsky, Prof. R.D. Adams, Dr. D. Chernous and K. Petrokovets.

REFERENCES

1.   Anfinogenov, S.B. Kurek, M.F. Shilko, & S.V. Chernous, D.A. (2008). Mechanical and frictional properties of biological elastomers: Part 1: Description of human skin relaxation under tension. Russian Journal of Biomechanics, Vol. 12, No. 3, pp. 42– 48, ISSN 1812-5123.

2.   Baughman, R.H. Galvao, D.S. (1993). Crystalline networks with unusual predicted mechanical and thermal properties. Nature, Vol. 365, pp. 735-737, ISSN 028-0836.

3. Bell, J.F. (1968). The physics of large deformation of crystalline solids. Springer tracts in natural philosophy. Vol. 14, Springer, Berlin-Heidelberg-New York.

4. Bergman, D.J. Inan, E. (Ed(s)). (2004). Continuum models and discrete systems, Kluwer Academic Publishers, ISBN 1-4020-2314-6, Dordrecht-Boston-London.Beverte, I.V., Kregers, A.F. (1987), Mechanics of Composite Materials. Vol. 23, No. 1, pp. 27-33).

5. Chernous, D.A. Shilko, S.V. Konyok, D.A. & Pleskachevsky Yu.M. (2003). Nonlinear viscoelastic behavior of flexible cellular plastics: refined rod model. International Journal of Applied Mechanics and Engineering, No. 1, pp. 27-41, ISSN 1425-1655.

6. Galaev, Yu.I. (1995). Smart polymers in biotechnology and medicine. Russian Chemical Reviews, Vol. 64, No. 5, pp. 505-524, ISSN 0036-021X.

7. Gibson, L.J. Ashby, M.F. (1982). The mechanics of three-dimensional cellular materials. Proceedings of the Royal Society A, Vol. 382, No. 3, pp. 43-59, ISSN 1364-5021.

8. Goldman, A.Ya. (1979). Strength of constructional plastics, Mashinostroenie, Leningrad (inRussian).

9. Hilyard, N.C. Cunningham, A. (1987). Low Density Cellular Plastics: Physical Basis of Behaviour, Chapman and Hall, London.

10. Hirotsu, S. (1991). Softening of bulk modulus and negative Poisson's ratio near the volume phase transition in polymer gels. Journal of Chemical Physics, Vol. 94, No. 5, pp. 3949-3957, ISSN 021-9606.

11. Kolupaev, B.S. Lipatov, Yu.S. Nikitchuk, V.I. Bordyuk, N.A. & Voloshin, O.M. (1996). Composite materials with negative Poisson coefficient. Journal of Engineering Physics and Thermophysics, Vol. 69, No. 5, pp. 542-549, ISSN 1062-0125.

12. Koniok, D.A. Voitsekhovsky, K.V. Pleskachevsky, Yu.M. & Shilko, S.V. (2004). Materials with negative Poisson's ratio. (The review). Journal on Composite Mechanics and Design, Vol. 10, No. 1, pp. 35-69, ISSN 1682-3532.

13. Lakes, R. (1987). Foam structure with a negative Poisson's ratio. Science, Vol. 235, pp. 1038- 1040, ISSN 0036-8075.

14. Landau, L.D. Lifshitz, E.M. (1986). Theory of Elasticity. Vol. 7 (3rd ed.), ButterworthHeinemann, ISBN 978-0-750-62633-0, Oxford.

15. Lederman, J.M. (1971). The prediction of the tensile properties of flexible foams. Journal of Applied Polymer Science, Vol. 15, No. 3, pp. 693-703, ISSN 0021-8995.

16. Prigogine, I. Stengers, I. (1984). Order out of Chaos: Man's new dialogue with nature, Flamingo, ISBN 0006541151, London.

17. Schwartz, M. (2007). Encyclopedia of Smart Materials, John Wiley & Sons, Inc. ISBN 0-471- 17780-6, New York.

18. Shilko, S.V. Stolyarov, A.I. (1996). Friction of anomalously elastic bodies. Negative Poisson's ratio. Part 2. Calculation of self-locking parameters. Journal of Friction and Wear, Vol. 17, No. 4, pp. 23–29, ISSN 0202-4977.

19. Shilko, S.V. Petrokovets, E.M. & Pleskachevsky, Yu.M. (2008). Peculiarities of friction in auxetic composites. Physica status solidi B, Vol. 245, No. 3, pp. 591–597, ISSN 0370- 1972.

20. Shilko, S.V. Petrokovets, E.M. & Pleskachevsky, Yu.M. (2008). Prediction of auxetic phenomena in nanoporomaterials. Physica status solidi B, Vol. 245, No. 11, pp. 2445–2453, ISSN 0370-1972.

21. Warren, W.E. Kraynik, A.M. (1987). The Winter Annual Meeting of the ASME, Boston, pp. 123–145.

22. Wei, G. Edvards, S.F. (1999). Effective elastic properties of composites of ellipsoids (II).Nearly disc– and nedle–like inclusions. Physica A, Vol. 264, pp. 404-423, ISSN 0378- 4371.

23. Wojciechowski, K. Alderson, A. Alderson, K.L. Maruszewski, B. & Scarpa, F. (2007). Preface. Physica status solidi B, Vol. 244, No. 3, pp. 813–816, ISSN 0370-1972.

Chapter 2

NETCENTRIC VIRTUAL LABORATORIES FOR COMPOSITE MATERIALS

E. Dado[1], E.A.B. Koenders[2], D.B.F. Carvalho[3]
[1]Netherlands Defence Academy, Breda, The Netherlands

[2]Delft University of Technology, Delft, The Netherlands COPPE-UFRJ, Programa de Engenharia Civil, Rio de Janeiro, Brazil

[3](PUC-Rio), Rio de Janeiro, Brazil

INTRODUCTION

Physical laboratory-based experiments and testing has been a way to develop fundamental research and learning knowledge for many areas of (civil) engineering education, science and practice. In the context of education, it has particularly enriched engineering education by helping students to understand fundamental principles and by supporting them to understand the link between the theoretical equations of their text books and real world applications. In the context of science, physical laboratory experiments have been used to scrutinize particular phenomenon in a real-life setting or to verify and validate scientific computational models over a longer period of time. In both the educational and research context, conducting physical laboratory experiments are generally governed by complex and expensive lab-infrastructures and require significant allocation of resources from the educational and research institutes. Besides, results most frequently have a limited range of exposure and are available for a relatively small audience (i.e. high costs versus relatively low benefits)

1]. In the context of practice, physical tests are often performed to validate performances of products. With the increasing regulations from national governments and European Union (EU) concerning quality, safety and environmental properties of products, the number of physical tests performed in laboratories of certificated (research) institutes have increased recently. For example, the European labels of conformity, known as CE markings, are a guarantee of quality and safety for products produced and sold in the EU.

According to the CE conformity standards, products will have undergone a series of performance tests before they can be sold on the EU market. However, these performance tests entail additional costs which may result in financial difficulties stemming from these additional costs and in the long run result in competitive disadvantages for Small and Medium Enterprises (SMEs) in Europe

2].To improve this situation, initiatives have been launched where research and development (R&D) projects could be conducted in so-called 'virtual laboratories'. In civil engineering, the National Institute of Standards and Technologies (NIST) in the United States were the first who set the standard for R&D projects conducted in a virtual laboratory for composite materials

3]. From this initiative and the promises of emerging information and communication technologies a whole new realm of possibilities for developing virtual laboratories has become available. Based on these observations and developments, the authors of this book chapter have initiated a number of R&D projects which main focus was to explore the concept of virtual laboratories for cement-based materials in real-life settings. A number of (journal) papers have been published about the findings of these projects in the past

4-7]. The main focus of this book chapter will be on a relatively new concept for establishing virtual laboratories which is based on a 'netcentric' approach.

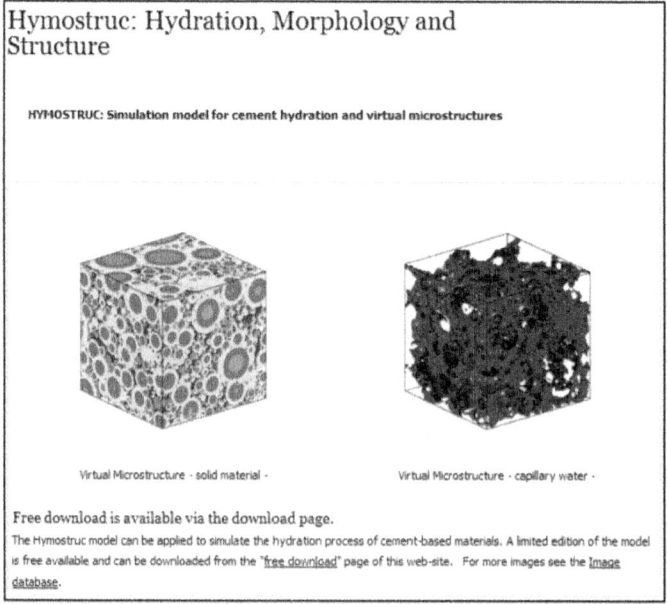

Figure 1: One of the results of the R&D projects conducted by the authors in the past.

THE CONCEPT OF NETCENTRIC VIRTUAL LABORATORIES

Virtual experiments (or testing) are rapidly emerging as a key technology in civil engineering. Although some applications of virtual experiments other than related to materials and components have been reported by a number of researchers, most effort has been put into the development of virtual laboratories for composite materials and components. In this respect, virtual experiments are often defined as a concept of making use of high performance computers in conjunction with high quality models to predict the properties and/or behavior of composite materials and components. Consequently, virtual laboratories are often seen as a new terminology for computer simulation, which is a wrong assumption. Although it is true that computer simulation is an important tool for virtual experiments, it is only one of key components that constitute a virtual laboratory. This can be explained by the limitation of the existing composite material models. As discussed by Garboczi et al, an ideal model of a composite material or structure should be one that starts from the known chemical composition of the composite material

8]. Beginning with the correct proportioning and arrangement of atoms, the modeling effort would build up the needed molecules, then the nanostructure and microstructure, and would eventually predict properties at the macroscale level. Such fundamental and multi-scale material model, however, is still a long way off. Each existing material model has its range of applicability and its own restrictions. Corresponding computational models and supporting computer tools are developed at a number of different research institutes worldwide. However, most existing virtual laboratories are setup as web applications that only provide access to a 'closed' virtual laboratory that contains a number of (integrated) computational models and supporting computer tools.

A good example of a closed virtual laboratory is the Virtual Cement and Concrete Testing Laboratory (VCCTL) from the National Institute of Standards and Technologies (NIST) in the United States. The main goal of the VCCTL project was to develop a virtual testing system, using a suite of integrated computational models for designing and testing cementbased materials in a virtual testing environment, which can accurately predict durability and service life based on detailed knowledge of starting materials, curing conditions, and environmental factors. In 2001, an early prototype (version 1.0) of the VCCTL became public available and accessible through the Internet. The core of this prototype was formed by the NIST 3D Cement Hydration and Microstructure Development Model (CEMHYD3D). Using the web-based interface of the VCCTL, one can create an initial microstructure containing cement, mine mineral admixtures, and inert fillers following a specific particle size distribution, hydrate the microstructure under a variety of curing conditions and evaluate the properties (e.g. chemical shrinkage, heat release, and temperature rise) of the simulated microstructures for direct comparison to experimental data. As the VCCTL project proceeded, the prediction of rheological properties (viscosity and yield stress) of the fresh materials and elastic properties (elastic modulus, creep, and relaxation) of the hardened materials were incorporated into the VCCTL resulting in the release of version 1.1 (latest) of the VCCTL in 2003

3].In order to cope with this particularity of distributed material (computational) models, the authors adopted a relatively new concept for establishing virtual laboratories which is based on a 'netcentric' approach. In this respect, a netcentric virtual laboratory is considered as a part of an evolutionary, complex community of people (users), devices, information (i.e. experimental data) and services (i.e. computational models and supporting computer tools) that are interconnected by the Internet. Optimal benefit of the available databases containing experimental data, computational models and supporting computer tools, that replace the physical laboratory equipment, is achieved via a distributed virtual laboratory environment and is assessable for students and researchers (see Figure 2).

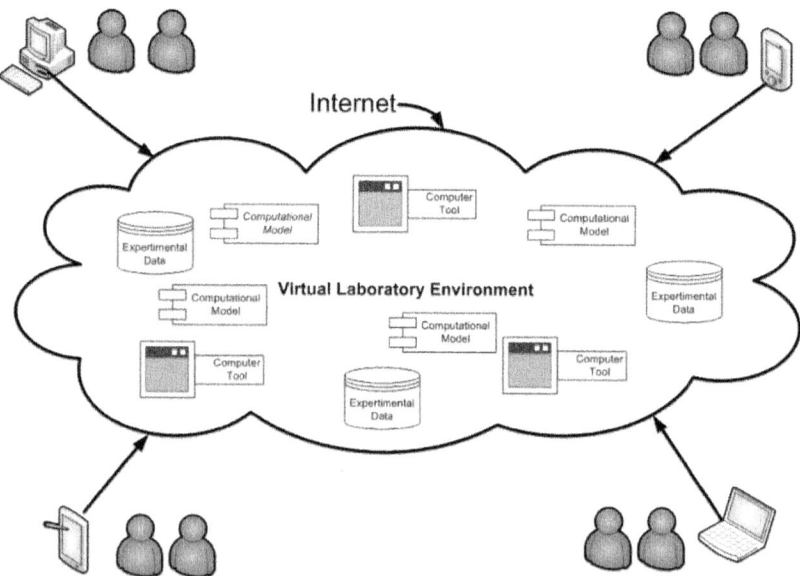

Figure 2: Virtual Laboratory Environment populated by devices, users, computational models and computer tools and databases with experimental data which are interconnected by the Internet.

Overview of Emerging and Enabling Technologies

As discussed in the previous paragraph, a virtual laboratory is no longer regarded as an isolated web-based application, but as a set of integrated devices (and supported infrastructures), computational models and supporting computer tools and databases containing experimental data that, used together, form a distributed and collaborative virtual laboratory environment for virtual experiments. Multiple, geographically dispersed (research) institutes will use this virtual laboratory environment to establish their own virtual laboratory to perform experiments as well as share their result from their R&D projects. As discussed in

5], emerging and enabling technologies should fundamentally concern about the integration (or interoperability), connecting devices, computational models, computer tools and data stores. In order to structure the discussion in this section, a conceptual scheme of the different levels of 'integration' is presented in Figure 3.

Integrated Infrastructure

Concerning the issue of the 'integrated infrastructure' two enabling and emerging technologies should be mentioned: Cloud Computing and Grid Computing. According to Foster et al.

9], Grid Computing and Cloud Computing are closely related paradigms that share a lot of commonality in their goals, architecture, and technology. According the National Institute of Standards and Technology (NIST) Cloud Computing (and to a large extend Grid Computing) can be defined as "a model for enabling convenient, on-demand

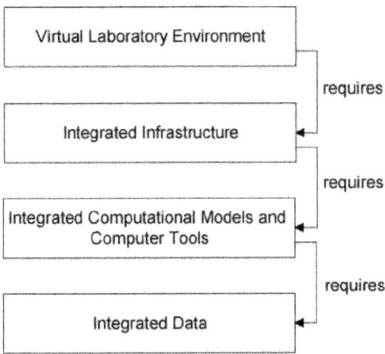

Figure 3: A Virtual Laboratory Environment requires an Integrated Infrastructure which on its turn requires respectively Integrated Computational Models and Computer Tools and Integrated Data.

network access to a shared pool of configurable computing resources (e.g., networks, servers, storage, applications, and services) that can be rapidly provisioned and released with minimal management effort or service provider interaction" [10]. In addition, Grid Computing technology adds the concept of 'virtual enterprise'. A virtual enterprise is a dynamic collection of institutes together in order to share hardware and software resources as they tackle common goals. As discussed in [3], a shared access to resources and the use of these resources is an inevitable condition for making multi-scale modeling successful. Cloud Computing and Grid Computing also share some limitations, namely the inability to provide intelligent and autonomous services, the incompetency to address the heterogeneity of systems and data, and the lack of machine-understandable content [11]. Mika and Tummarello (2008) identified the root cause of these limitations as the lack of 'Web semantics' [12]. These limitations should be addressed at service and data levels as discussed in the next sections.

Integrated Computational Models & Computer Tools

Traditionally, the programming environment at research institutes is dominated by (programming) languages such SQL for storing, retrieving and manipulating data, Fortran, C, and C++ and Java for implementing computational models and HTML for developing web-based interfaces for end-users. Web Services fundamentally concern about the integration of computer programs, especially when the computer programs concerned are developed using different programming languages and computer operating platforms. Web Services standards and technologies offer a widely adopted mechanism for making computer programs work together.

Currently, the two main players on the web services market are Oracle (by the ownership of Sun), with their Java platform and Microsoft with the .NET platform, where both agree on the core standards (e.g. SOAP, WSDL, UDDI and XML), but disagree on how to deliver the potential benefits of Web Services to their customers. Simple Object Access Protocol (SOAP) is the standard for web services messages. Based on XML, SOAP let web services exchange information over HTTP. In addition, WSDL (Web Services Description Language) is an XML-based language for describing web services and the way how to access them. UDDI (Universal Description, Discovery, and Integration) is an XML-based registry for web services to list themselves on the Internet. XML (EXtensible Markup Language) has replaced HTML as de defacto standard for describing defining and sharing data on the Internet. The most important advantages of XML are: (1) its separation of definition (content) and representation (mark-up), and (2) its ability to support the development and use of domain specific XML vocabularies. Using the web service standards SOAP, WSDL and UDDI make computational models and supporting computer tools web services that can be accessed, described and discovered. Using XML as a basis for sharing data on the Internet will solve the interoperability problems at data level as described in the next section.

Integrated Data

As discussed earlier, one of the main causes that Cloud Computing and Grid Computing share some limitations is the lack of 'web semantics'. Using XML will not solve the problem entirely; it has also its own limitations. The limitations of XML were solved by the introduction of the Semantic Web in 2004. Driving the 'Semantic Web' is the organization of content specialized vocabularies, referred to as 'ontologies'. In this respect, ontology is a collection of concepts (or terms) and constructs used to describe a particular knowledge domain. Build upon RDF (Resource Description Framework) and XML and derived from DAML/OIL, OWL (Web Ontology Language) has become the default standard for creating ontologies. Building ontologies for describing and exchanging data between computational models and supporting computer tools in virtual laboratory environment will result in a number of different ontologies. In this respect, three different types of ontologies can be distinguished: (1) high-level (reference) ontologies that hold common concepts which can be applied for all knowledge domains and hold high-level constructs for defining the relationships between these knowledge domains, (2) knowledge domain (reference) ontologies which hold the concepts and constructs that are common within one specific knowledge domain (i.e. referring to the different macro-, meso-, micro- and nano-scale levels that exist in material research) and (3) application ontologies that hold detailed information about concepts and constructs which form the basis for sharing data between a 'group' of computational models and supporting computer tools on the Internet. From a modeling point of view, each application ontology is an extension of one or more knowledge domain ontologies, while each knowledge domain ontology is an extension of one or more high-level ontologies. Together, they form a so-called 'ontology network' that evolutionary changes over time.

A MULTI-SCALE MODELING APPROACH FOR CONCRETE MATERIALS

The development of a virtual laboratory for construction materials requires many years of research to find out the basic principles of which a virtual laboratory should comply with, and also to find out the conditions at which a virtual laboratory would be attractive to researchers, students and people from the industry

13]. At the Delft University of Technology, first trials were focused on the virtual testing of a concrete compressive strength test, where the hydration conditions and the fracture behavior where evaluated with emphasis on the upscaling of the simulation model results. In physical-based concrete laboratories, the compressive strength is determined with an experimental device where a concrete cube is positioned in between two steel plates and compressed using a hydraulic force (Figure 4, left). The force imposed to the concrete cube is increased until failure occurs. Later, the focus was widened to other concrete properties such as tensile strength (Figure 4, right), elastic modulus, hydration temperature, etc.

Figure 4: Experimental testing device for concrete compression (Left) and tension (Right).

In a virtual laboratory, testing procedures and methods have to be mimicked using computer simulation models. These computer models are applied to simulate the different characteristics of cement-based materials and produce results that may be validated with experimental data. Especially for heterogeneous materials such as concrete materials, characteristics are modeled at different scale-levels requiring the modeling approach to be multi-scale. This means that models, operating at scale length, have to communicate with each other and exchange information by means of parameters passing which require upscaling algorithms. In general the following scale levels can be distinguished that associate with the characterization of specific materials properties (Table 1).

Table 1: Overview of modeling scales, properties and size ranges

Scale	Property	Size range [m]
Macro	Rheology / Mechanical / Cracking / Volume stability / Durability	$10^{-1} - 10^2$
Meso	Compressive and Tensile strength / Fracture energy and Toughness	$10^{-5} - 10^{-1}$
Micro	Hydration / Chemistry / Pore pressure / Permeability	$10^{-6} - 10^{-5}$
Nano	C-S-H analysis, Calcium leaching / CH, Ca/Si -ratio / Al/Si – ratio	$10^{-10} - 10^{-6}$

The input for the different scale level models consists either of direct user input or of input achieved from a lower or higher scale level. This will lead to a refinement of the simulation predictability and will lead to a system with an increased synergy. A virtual laboratory can help to facilitate this linking of models and provides the opportunity to allow other partners to link their models as well, leading to an overall modeling platform for computational materials design (see Figure 5).

Figure 5: Schematic representation of the modeling levels.

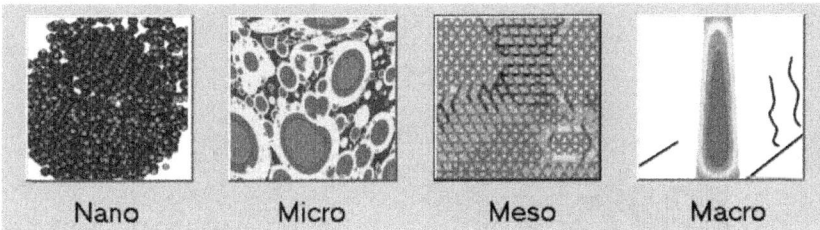

Nano   Micro   Meso   Macro

Macro-scale level

The macro-scale level is the level at which full-scale structures are being designed and calculated. Relevant concrete related issues that have to be known for this particular scale level are listed in Table 1. For the macro-scale level, Finite element models (FEM) are very often used to simulate the structural respond of systems under static and/or dynamic actions, but also to simulate the early age hardening behavior of freshly cast material. Simulation models are used to avoid cracking during hardening. Practical problems are often related to the mix design and to the climatic conditions under which hardening takes place. Design and engineering of concrete structures, therefore, can be conducted with macro-scale FEM models like DIANA [14], ANSYS [15], ABAQUS [16] or FEMMASSE [17], where the latter is especially designed for early age analysis of concrete structures. In particular when considering the early age behavior, the development of the materials properties becomes very relevant. FEM models need to receive this information as input, or sometimes simple mathematical formulas or empirical functions are developed that can be fitted to experimental data and, with this, are able to predict the materials behavior. The disadvantage of this approach is that still expensive and time consuming physical laboratory experiments have to be conducted for the fitting and validation of the models. With these FEM models the stress and strength development can be calculated for full-scale concrete structures during the early stage of hardening. For accurate assessments, the model requires input from the structure's geometry and formwork characteristics, the ambient conditions and the thermo-mechanical properties of the hardening mix. From these inputs, the model calculates the temperature field and, from this, the development of the tensile stresses that occur as a result of the internal and external restraint (Figure 6). Important material properties herein are the development of the tensile strength, the elastic modules and the relaxation coefficient. All these properties are depending on the mix composition and on the hardening conditions, and can be expressed as a function of the degree of hydration.

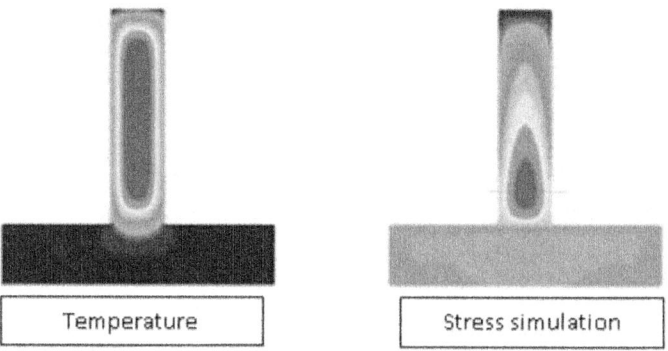

Temperature   Stress simulation

Figure 6: Temperature and stress field during early hydration of a hardening concrete wall cast on an already hardened slab [17].

With the calculated stress field and the tensile strength development known at any location in a structure, the probability of (macro) crack occurrence can be calculated using statistical methods. Cracking will occur if the calculated stress exceeds the strength while accounting for the scatter of the material properties. The accuracy of the crack predictions also requires an accurate description of the material properties. Experienced-based models, databases or numerical simulation models operating at a more detailed scale level can be used for this. More fundamental models, based on thermo-mechanical-physical mechanisms operating at an increased level of detail, can also be applied. In the next sections, models operating at the meso, micro and nanoscale level will be discussed with emphasis on the increased level of detail for the simulations while aiming to improve the prediction accuracy of the models that operate at the macro-scale level.

Meso-scale level

Increasing the level of detail of numerical schemes with the objective to simulate fracture propagation processes in concrete, lattice models can be applied. In order to be able to predict the ultimate capacity of a virtual concrete sample by simulating its cracking pattern, a fracture mechanics model is required which can handle the crack propagation of the material while loaded. A model that can be adopted in this respect is the Delft Lattice model, which has originally been developed by Schlangen and Van Mier in 1991

The model simulates fracture processes by means of mapping a framework of beams to a materials meso structure. The basic principles of the Lattice model are schematically shown in Figure 7. In this figure (a) is a schematization of a regular type of framework mesh that can be used to simulate the meso-level structure of brittle materials, such as concrete.

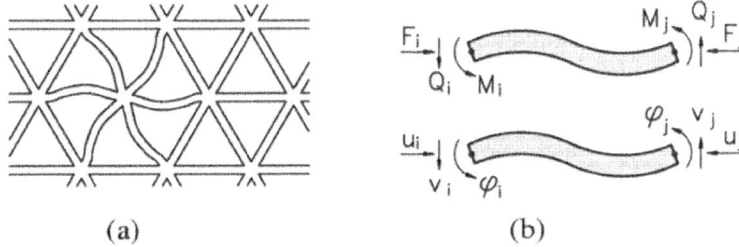

(a)   (b)

Figure 7: Principle of the Lattice model [18]; a) Lattice framework, b) Lattice beams with action forces and displacements indicated.

For composite materials in particular, the meso-level structure reflects the schematization of the paste phase, the Interfacial Transition Zone (ITZ) and the aggregate explicitly. This approach of schematization fits very well with the level that is required for modeling the compressive stress calculations inside a virtual laboratory. The model should be able to detect failure paths through the material (weakest links) and to calculate the accompanying ultimate strength of the building material from it. Once this failure path has been initiated, the inner structure of the material starts to disintegrate and the strength capacity will reach its maximum. After having reached this maximum strength level, a descending branch will follow that indicates the post-peak behavior of the material. The Lattice model is capable to calculate this part of the failure traject and to quantify the fracture energy of the failure behavior as well. For conventional concretes, the Interfacial Transition Zone (ITZ, weak bonding zone around the aggregates) is almost always the weakest part of the material that initiates and contributes to the failure paths (Figure 8). For higher quality concretes, the failure paths might cross through the aggregate particles which implicitly affect the brittleness of the material. A proper compressive strength model within a Virtual Laboratory should therefore implicitly deal with these different kinds of failure mechanisms related to the mix composition in general and the inner microstructure of the material in particular.

Micro-scale Level

For the simulation of the evolving cementitious microstructure that forms the fundamental basis for the development of the material properties the hydration model Hymostruc can be used [19,20]. After mixing, hardening commences and the material properties start to develop. This process leads to a set of properties that is unique for every particular type of cement-based material. The Hymostruc model (Figure 9, left) can be applied to predict the actual state of the material properties with the degree of hydration as the basic parameter. The model calculates the hardening process of cement-based materials as a function of the water-cement ratio, the reaction temperature, the chemical cement composition and the particle size distribution of the cement. The model calculates the inter-particle contacts by means of the 'interaction mechanism for the expanding particles' (Figure 9, right) where hydrating particles are embedded in the outer shell of larger hydrating particles. This mechanism provides the basis of the formation of a virtual microstructure which, on its turn, can be considered as the backbone of the evolving strength capacity of the material. When considering the virtual laboratory, the Hymostruc model will operate at the micro-scale level and will be used to calculate the internal microstructure that is necessary to simulate the compressive and tensile strength development, the development of the elastic modulus and other microstructure related properties. The microstructure of the material can be considered as the morphology-based inner structure of the paste, i.e. the "glue", that tightens together the aggregate particles and/or other composite phases such as fibers, fillers, etc, inside a composite material. Failure of the paste structure, therefore, strongly depends on the strength characteristics of the internal bondings in the microstructure of the paste. Modeling the morphologies of these bondings in terms of their chemo-physical nature has to be resolved at the nano-scale level. The development of the properties of the C-S-H gel, therefore, is a scale-level that has to be considered as well.

 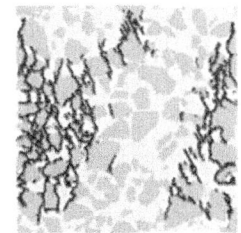 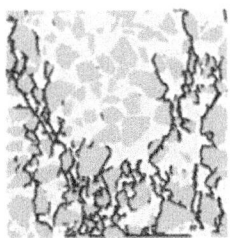

Figure 8: Left: Concrete crack pattern after loading. Right: Lattice simulation [18].

Nano-scale Level

Nano-scale modeling has benefit from an enormous increase in the attention of the research community with the aim to model the chemical and physical based processes of the Calcium-Silicate-Hydrates (C-S-H gel) that forms the fundamental elements of the hydration products of cementitious materials [21]. Characterizing the materials performance at this particular scale level asks for modeling the fundamental processes using molecular dynamics principles. For cement-based materials in particular, emphasis has to be on the characterization of the basic building blocks of the C-S-H nano-structure that operate at the sub-micro scale level. This intermediate scale-level between the nano and micro-scale level enables a modeling approach that bridges the gap between the micro and nano level and that enables an exchange of fundamental materials properties (Figure 10).

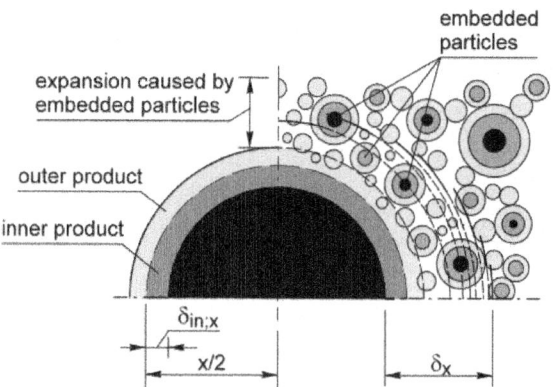

Figure 9: Left: 3D virtual microstructure simulated with Hymostruc. Right: Hymo-struc interaction mechanism for expanding particles representing the formation of structure of the virtual microstructure [19,20].

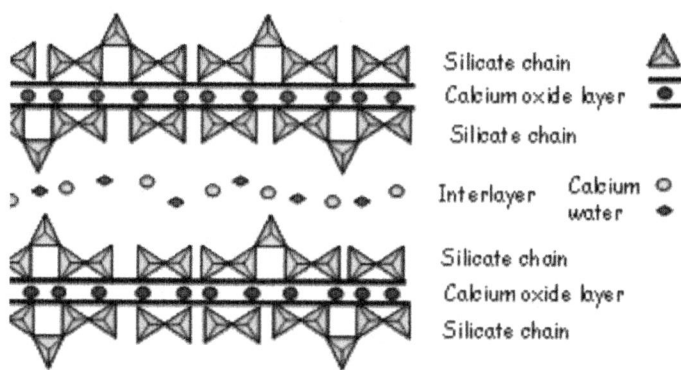

Figure 10: Structural model that describes the atomic scale of the C-S-H gel [21].

Up-scaling

The development of numerical algorithms that allows for particular scale-level information to be used at other scale levels is the most challenging part of multi-scale modeling. Bridging the length scales between the nano scale level and the macro-scale level requires a upscale models that enable to span of 10 orders of magnitude (see Table 1). In this approach the nano scale level forms the basis of the multi-scale framework. The output properties calculated at a particular scale level forms the input at a higher scale level. This approach enables the analysis and design of composite materials starting from the fundamental nanoscale level and evaluates the results at the full-scale macro-level. It opens the door for tailor made design of composite materials and is a first step towards property defined modeling approach. A virtual laboratory is an excellent vehicle to achieve this.

Virtual Laboratory Prototype

In 2012, the first prototype of a virtual laboratory has been developed at Delft University of Technology. In this paragraph, the system development rationale and architecture are presented, including screenshots of the prototype created to demonstrate the idea of a webbased virtual laboratory. Due to the fact that existing computation models and supporting computer tools have been implemented in the traditional way in the past, the demonstrated prototype not fully operates according the netcentric approach as discussed above. The main purpose of this prototype is to support the approach of multi-scale modeling for concrete materials in an integrated web-based environment. From this, it can be derived that prototype should provide a complete environment for multi-scale experimentation easing the study of composite materials.

From the point of view of the final users, the prototype should aid and support them on their experiments relying on a fast execution of the best simulation models existing. Since the system's user interface is based on the functionalities available in the computational (simulation) models and supporting computer tools, the final user needs to know about the computational models to be able to work with them. However, this user does not want to care where these computational models are running – they want to focus on their educational and research applications instead of the technology required providing their needs. For instance, the system should relief the users from the burden of looking for simulation module availability, its installation and execution; take care of the issues regarding combining modules to perform multi-scale experiments and carry out other system administration duties too. From the point of view of the computational model creators, the virtual laboratory should be open in a way to be able to support new computational models to be plugged-in adding new services and functionalities to the existent ones. The computational models must exchange data, making possible the execution of a multi-scale experiment based on different computational models developed independently (i.e. mashup). Therefore, there must exist a platform that works as an open ecosystem environment, populated with simulation models, created and executed independently, but that are coordinated by the final users through an interface that offers multi-scale experiments of composite materials. With this purpose, the system architecture was created based on two principal modules, the user interface module (frontend) and the simulation modules (backend). This modular organization provides decoupling of the user interface from the simulation modules. The deployment diagram of the virtual laboratory system is presented in Figure 11. showing the fronted module as the Graphical User Interface (GUI) component and the backend module as the model execution controller, which relies on a Grid infrastructure to execute the computational models.

The virtual modelling laboratory contains a rich interface web application that has been developed using cutting-edge technology, such as HTML5, CSS3 and JavaScript. The backend is defined as a set of web services which are developed as a Restful API and which is accessed through AJAX calls. An extensive set of toolkits and frameworks is used in the final implementation. The GUI foundation is based on the Twitter Bootstrap library for UI building and on the Backbone.js toolkit for organizing the JavaScript code, and being fully based on the jQuery library. The backend was developed using the Play framework. Although the basic application architecture is defined as a two-layered system – frontend and backend, the backend layer is composed by a set of different computational modules that can be seen as independent web services. The modules executions are managed coordinated by the overall controller service, but the composition architecture is much more complex than what is exposed. Different researchers and developers, distributed over the Internet, can make their computational models available through different interfaces. In addition, the computational models can also use results from other computational models.

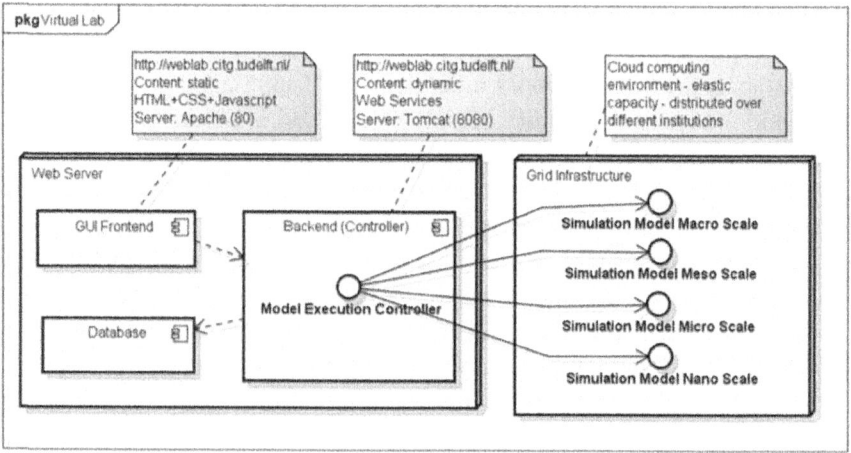

Figure 11: The deployment diagram of the virtual laboratory at Delft University of Technology.

As discussed earlier, the prototype relies on a Grid infrastructure to execute the computational models. This infrastructure is distributed over different institutions, because each computational model owner should be responsible for the model execution and availability as a web service. For this purpose, a Grid infrastructure should be available to support the whole platform. As discussed earlier, other new emerging technologies such as Cloud Computing can supply on-demand computational resources to run the computational models. In order to be able to transfer data between the different modelling scale levels, multi-scale modelling principles have to be considered as well. Apart from how the data is transferred from one scale level to another, the most challenging part is how to connect the levels from a modelling point of view. Bridging the scale levels can go along with transfer of data only by means of parameters passing or by means of a more complicated integration of models that operate at different scale levels or a possible combination. In either way, bridging the scale levels is an intensive modelling work that requires significant effort both from materials properties as well as from a modelling point of view. With the virtual modelling lab, the user can choose at which level he/she wants to start the numerical experiments (Figure 12). It is organized in such a way that data can be calculated in a certain scale level and, when the user decides to proceed with an analysis at another scale level, the data can be taken.

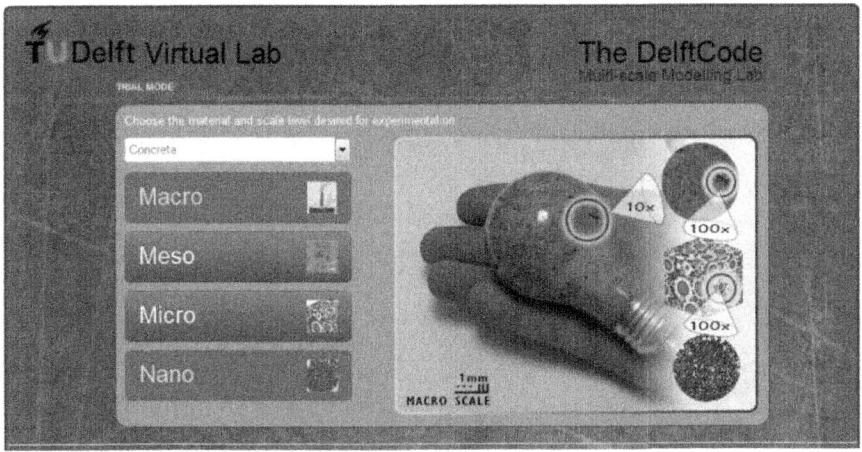

Figure 12: DelftCode: Selection of scale level.

The pilot version of the virtual laboratory at the Delft University of Technology (referred to as DelftCode) provides output (results) which are represented as graphs in the GUI. In addition, each simulation of computational model at a certain modelling scale produces specific parameter output that is managed by the DelftCode framework in such a way that it can be used as input for computational models at other scale levels (Figure 13). In this way active multi-scale modelling can be conducted and the results can be reused at other scale levels. The way how it is implemented in the DelftCode framework is that after conducting numerical experiments the user can switch to other scale levels. For the other scale levels the same procedure is followed. Since data will be stored and available for all scale levels, the user can access data of previous numerical experiments and reuse it for other experiments.

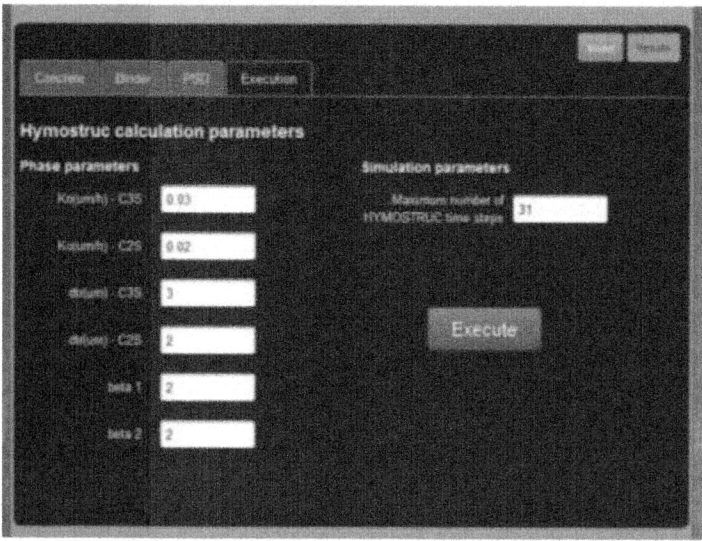

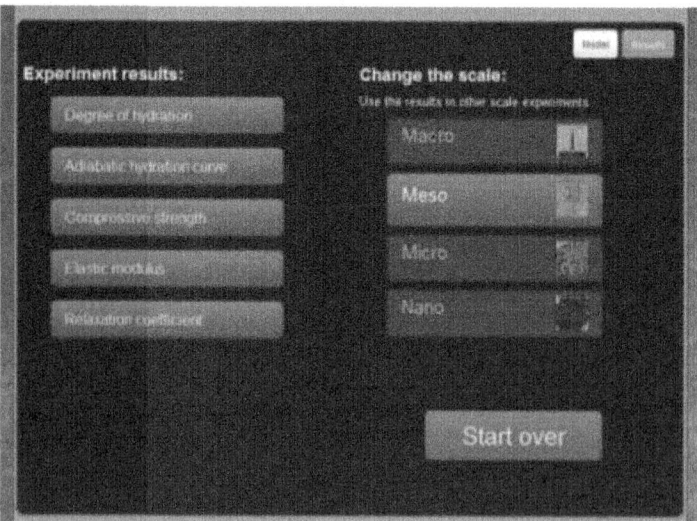

Figure 13: DelftCode: Model setup and result outputs.

At time of writing, the prototype only supports computational models at micro and meso scales, but the prototype has been designed and implemented to support all scale levels. Figure 14 show all the proposed user interactions in the prototype. These interactions show the possibilities of the developed prototype that allows multi-scale modeling platform. The architecture of the prototype is designed in such a way that future extensions in terms of new model additions or adding another scale level can easily be achieved.

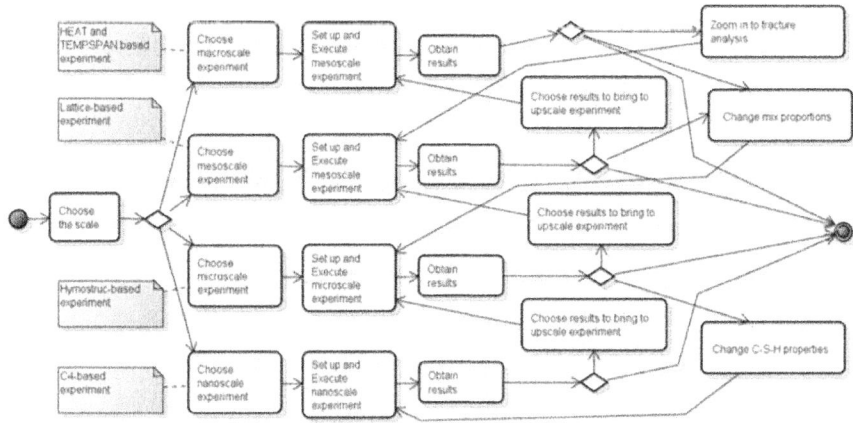

Figure 14. The activity diagram of the user experiments of all scales.

CONCLUSION AND DISCUSSION

This paper depicts the development the concept of so-called virtual laboratories that are based on a netcentric approach. In this context, a netcentric virtual laboratory is considered as a part of an evolutionary, complex community of people (users), devices, information (i.e. experimental data) and services (computational models and supporting computer tools) that are interconnected by the Internet. A large number of emerging and enabling technologies are discussed which form technological basis for establishing such netcentric virtual labs. Although the developed prototype is hampered by the fact that most computational models and supporting computer tools have been developed using traditional programming languages and platforms, it showed the enormous potential of the netcentric approach advocated in this paper. From the perspective of the application domain, a virtual laboratory can be considered as a most appropriate way to interact with users and developers of computational material models at different scale levels. In this paper the approach of multi-scale modeling approach for concrete materials is explained. This approach is based on numerical (computational) models developed for cementitious materials that operate at different 'geometrical' scale levels. With the ability to use the output generated at any particular scale level as input for models that run at other scale levels, the web-based virtual laboratory acts as a real multi-scale modeling platform. The architecture of the proposed virtual laboratory as provided in this chapter allows the models to be exchangeable and merge-able leading to an integrated approach.

The numerical models for simulating the materials performance operate at the different scale-levels with the Hymostruc model as the main microstructural model at the micro-scale level. With this model connecting to the nano-scale model for inputting detailed information on C-S-H gel properties, the microstructural information can be used as input for the mesolevel Lattice model to simulate the fracture behavior of composite materials submitted to internal actions (drying, autogenous shrinkage, etc) or external actions (loads, thermal imposed loading, etc). These models can generate input data for the macro-scale models to simulate the full scale performance of structural elements. With the multi-scale approach, the consequences of changing parameters that act as input for lower scale models (nano, micro, meso) can directly be made visible by upscaling. Therefore, from this approach the following conclusions can be drawn:

- The netcentric virtual laboratory is a most appropriate tool for the assessment of composite materials performance using a multi-scale modeling approach;
- The web-based approach enables the communication between models that operate at different geometrical scale-levels using an integrated computational modeling system;
- The prototype shows the huge potential of web-based modeling and provides an exchangeable and scalable system for multi-scale modeling;
- The future perspective of virtual web-based modeling shows to be a very powerful alternative for vast computational models that run over different time and length scales.

RTEFERENCES

1. Kuester, F. and Hutchinson, T. A virtualized laboratory for earthquake engineering education, ASEE Journal of Engineering Education, 15:1, 2007.

2. ECWINS consortium. The Road to Standardized Window Production. Collective Research Projects for SMEs, Vol. 3, European Union, 2007.

3. Bullard, J. et al. Virtual Cement, Innovations in Portland Cement Manufacturing, USA, 2004.

4. Dado, E., Koenders, E. and Mevissen, S. Towards an advanced virtual testing environment for concrete materials, published in the proceedings of the MS2010 conference, 2010.

5. Dado, E, Koenders, E. and Beheshti, R. Theory and Applications of Virtual Testing Environments in Civil Engineering, International Journal of Design Sciences and Technology, 16:2, 2009.

6. Koenders, E., Schlangen, E. and Dado, E. Virtual testing of compressive strength of concrete, published in the proceedings of the ISEC-4, Conference, 2007.

7. Koenders, E., Dado, E and van Breugel, K. A Virtual Environment for Multi-Aspect Modeling, published in the proceedings of the SCI2004 conference, 2004.

8. Garboczi, E., Bullard, J. and Bentz, D. Virtual Testing of Cement and Concrete, Concrete International, No. 12, United States, 2004.

9. Foster, I., Zhao, Y., Raicu, I. and Lu, S. Cloud Computing and Grid Computing 360- Degree Compared. Proc. Grid Computing Environments Workshop, p.1-10, 2008.

10. Mell, P. and Grance, T. The NIST Definition of Cloud Computing, NIST special publication 800-145, United States, 2011.

11. Wu, Z. and Chen, H. From Semantic Grid to Knowledge Service Cloud, Journal of Zhejiang University, 13:4, China, 2012.

12. Mika, P. and Tummarello, G. Web semantics in the clouds. IEEE Intelligent Systems Magazine, 23-5, United States, 2008.

13. Koenders, E, Schlangen, E. and Dado, E. Virtual Testing of Compressive Strength of Concrete, Proceedings of the ISEC-4 Conference, Australia, 2007.

14. DIANA, http://tnodiana.com/.

15. ANSYS, www.ansys.com/.

16. ABAQUS, www.simulia.com.

17. FEMMASSE, www.femmasse.nl.

18. Schlangen, E. Experimental and Numerical Analysis of Fracture Processes in Concrete, PhD thesis, Delft University of Technology, The Netherlands, 1993.

19. Breugel, van, K. Simulation of Hydration and Formation of Structure in Hardening Cement-based Materials, PhD. Thesis, Delft University of Technology, Delft, The Netherlands, 1991.

20. Koenders E. Simulation of Volume Changes in Hardening Cement-Based Materials, PhD-Thesis, Delft University of Technology, Delft, The Netherlands, 1997.

21. Dolado J. S., Hamaekers J. and Griebel M. A Molecular Dynamic study of cementitious Calcium Silicate Hydrate (C-S-H) gels, Journal of American Ceramic Society.

Chapter 3

FREQUENCY-DEPENDENT EFFECTIVE MATERIAL PARAMETERS OF COMPOSITES AS A FUNCTION OF INCLUSION SHAPE

Konstantin N.[1], Marina Y. Koledintseva[2],and Eugene P. Yelsukov[3]
[1]Institute for Theoretical and Applied Electromagnetics, Russian Academy of Sci., Moscow, Russia

[2]Missouri University of Science and Technology, Rolla, MO, USA

[3]Physical and Technical Institute, Ural Branch of Russian Academy of Sci., Izhevsk, Russia

INTRODUCTION

Electromagnetic composite materials have a number of promising applications in various radio-frequency (RF), microwave and high-speed digital electronic devices, and allow for solving problems related to electromagnetic compatibility (EMC) and electromagnetic immunity (EMI) [1]. For this reason, study and prediction of frequency-dependent radio-frequency RF and microwave properties of materials currently attract much attention. The problem of interest is the analytical description of wideband RF/microwave permittivity and permeability behavior of materials. This is necessary, in particular, to numerically optimize wideband electromagnetic performance of materials and devices at the design stage.

This chapter discusses frequency dependences of effective material parameters (permittivity and permeability) of different types of composites. The chapter consists of three sections. Section I presents a review of approaches for predicting effective material parameters of composites, such as mixing rules, the Bergman–Milton spectral theory, and the percolation theory. Section II suggests on how to select the most appropriate mixing rule for the analysis of properties of a particular composite. Section III considers the dielectric microwave properties of composites filled with fiber-shaped inclusions.

APPROACHES TO DESCRIBE EFFECTIVE MATERIAL PARAMETERS OF COMPOSITES

Basic Mixing Rules

In most studies, two-component mixtures are considered, where identical inclusions are imbedded in a homogeneous host matrix. Effective properties of such a composite depend on the intrinsic properties of the inclusions and the host matrix, as well as on the morphology of the composite. The morphology is a characterization of the manner, in which inclusions are distributed in the composite, including their concentration, shape, and correlations in the location. Therefore, the morphology determines how inclusions are shaped and distributed, whether they are mutually aligned/misaligned in the composite, and what concentrations of inclusion phases and a matrix material are.

A conventional approach to describe the properties of composites employs mixing rules, i.e., equations that relates the intrinsic properties of inclusions and the host matrix with the effective properties of composite based on a simple idealized model considering an ellipsoidal-shaped inclusion. Typically, the characterization of the concentration and the shape of inclusions are included explicitly in the mixing rules, and the account for other morphological characteristics is attempted by a proper selection of the mathematical form of mixing rules.

A number of mixing rules are found in the literature. The basic mixing rules are the Maxwell Garnet equation (MG) [2], Bruggeman's Effective Medium Theory (EMT) [3], and the Landau-Lifshitz-Looyenga mixing rule (LLL) [4, 5]. The MG mixing rule,

$$\frac{\chi_{eff}}{1 + n\,\chi_{eff}} = p\,\frac{\chi_{incl}}{1 + n\,\chi_{incl}},$$

$$(1)$$

is equivalent to the Clausius–Mossotti approximation, and also complies with the Ewald-Oseen extinction theorem [6]. Bruggeman's EMT,

$$-(1-p)\frac{-\chi_{eff}}{\chi_{eff} + 1 - n\chi_{eff}} = p\,\frac{\chi_{incl} - \chi_{eff}}{\chi_{eff} + 1 + n\left(\chi_{incl} - \chi_{eff}\right)},$$

$$(2)$$

is often referred to as the Polder-van Santen mixing rule in the theory of magnetic composites [7]. The Landau–Lifshitz–Looyenga mixing rule (LLL) is written as

$$\left(\chi_{eff} + 1\right)^{1/3} - 1 = p\left(\left(\chi_{incl} + 1\right)^{1/3} - 1\right)$$

$$(3)$$

Eqs. (1)–(3) are written for the generalized susceptibilities of inclusions, χ_{incl}, and the effective susceptibility, χ_{eff}, both normalized to the susceptibility of the host matrix, since all susceptibilities of a certain composite, including the effective permittivity and permeability, are governed by the same mixing rule [8], with a possible correction for the tensor nature of the susceptibilities. If permittivity $\varepsilon=\varepsilon'-i\varepsilon''$ is under consideration, then $\chi_{incl}=\varepsilon_{incl}/\varepsilon_{host}-1$ and $\chi_{eff}=\varepsilon_{eff}/\varepsilon_{host}-1$. For the permeability $\mu=\mu'-i\mu''$, $\chi_{incl}=\mu_{incl}-1$ and $\chi_{eff}=\mu_{eff}-1$, because most magnetic composites are based on a non-magnetic host matrix. In Eqs. (1)–(3), n is the form factor, i.e., either depolarization or demagnetization factor, and p is the volume fraction of inclusions.

Starting from the basic mixing rules, simple empirical models of a composite may be suggested.

The MG mixing rule considers the total polarizability of inclusions represented by the right part of Eq.(1) and assumes that this polarizability is acquired to a homogeneous medium. As a consequence, this mixing rule defines the weakest possible cooperative phenomena between neighboring inclusions that are feasible for the given volume fraction of inclusions. The MG mixing rule is an accurate result for the case, when excitation of inhomogeneous fields due to multiple scattering on inclusions and the effect of neighboring inclusions are negligible. Therefore, it coincides with the lower Hashin–Shtrikman limit [9] that provides the smallest possible material parameter in the case, when loss is negligible. The MG mixing rule is believed to be valid for regular composites, i.e., those comprising regularly arranged inclusions, and for the case of conducting inclusions covered with an isolating shell [10, 11].

The physical model for the EMT assumes that the host matrix consists of particles having the same shape that inclusions have, and both the inclusions and the host matrix particles, are embedded in an effective medium with the material constant equal to the effective material constant the composite. The sum of the polarizabilities of these two types of particles must be zero, which corresponds to a homogeneous medium. Though a practical realization of the EMT involves a very special morphology of a composite [12], this mixing rule is widely used, because it incorporates the percolation threshold when modeling a metal-dielectric mixture. The percolation threshold, p_c, is the lowest concentration, at which a macroscopic conductivity appears in the mixture. From the standpoint of the mathematics, the EMT is reduced to a quadratic equation for the effective permittivity. Below and above the percolation threshold, different solutions of the equation must be selected according to the physical selection rules. Equation (2) yields $p_c=n$. In the MG mixing rule (1), the percolation threshold is $p_c=1$.

The LLL mixing rule is built up by an iterative procedure starting from a homogeneous material of inclusions and replacing small amount of this material by the material of the host matrix. After that, the resulting "effective" material is regarded as the homogeneous component for the succeeding substitution step, and so on, which results in Eq. (3). The mixing rule obtained by the same iterative procedure starting with the homogeneous host matrix is referred to as the asymmetric Bruggeman approximation. The result of the LLL mixing rule is independent of the form factor of inclusions. For a metal-dielectric composite, the LLL mixing rule always provides a conductive mixture, so that $p_c=0$.

The LLL mixing rule is known to be an accurate result for the case when the material parameter of inclusions differs slightly from that of the host matrix. In particular, this mixing rule is valid for all material parameters of composites at very high frequencies, because any intrinsic susceptibility of any material approaches zero with the frequency tending to infinity. Agreement of both the MG and EMT mixing rules with the LLL mixing rule in the case of the susceptibility of inclusions slightly differing from zero is attained only if $n=1/3$.

When the volume fraction of inclusions is small, $p \ll p_c$, and the interaction between the inclusions is negligible, all three theories are reduced to the small perturbation limit,

$$\chi_{\text{eff}} = \frac{p\chi_{\text{incl}}}{1+n\chi_{\text{incl}}}$$

(4)

Strictly speaking, Eqs. (1), (2), and (4) are valid for the case of inclusions of perfectly spherical shape, which have the shape factor equal to 1/3. Otherwise, the equations are not consistent with the LLL mixing rule in the limiting case of high frequencies. For non-spherical particles, the polarizability of inclusions must be averaged over all three principal axes of the inclusion [13]. Two particular cases of non-spherical inclusions are of practical interest - nearly spherical inclusions and highly elongated inclusions (long fibers or platelets). For nearly spherical inclusions, the composites are conventionally described by Eqs. (1), (2), (4) with averaged form factor involved, which is found empirically and may differ from 1/3. For elongated inclusions, the form factor along the shorter axis (in the platelet case), or the sum of two form factors along the shorter axes (in the fiber case) is close to unity, and the polarization of inclusion is these directions can be neglected. In this case, the above equations are valid again, with a randomization factor, κ, included in the right-hand part of the equations to account for an alignment of non-spherical inclusions. For a fiber-filled composite, $\kappa=1/3$, when the fibers are randomly oriented in space, and $\kappa=1/2$, when the fibers are randomly oriented in plane and the wave vector in perpendicular to the plane. For composites filled with platelet-shaped inclusions, $\kappa=2/3$ for the 3D isotropic orientation.

For the permeability of composites filled with non-spherical particles, possible anisotropy of magnetic moment, associated with crystallographic anisotropy of particle material, should be also taken into account, see, e.g., [14].

The Bergman-Milton Theory

A generalization of mixing rules may be made with the use of the Bergman-Milton spectral theory (BM) [8]. The theory expresses the effective material parameter of a composite as

$$\chi_{\text{eff}} = p \int_0^1 \frac{\chi_{\text{incl}} b(n) \, dn}{1 + n \chi_{\text{incl}}} \tag{5}$$

where the spectral function, $b(n)$, is introduced as a quantitative characterization of the composite morphology. As is seen from Eq. (5), the BM theory accounts for a distribution in effective form factors of inclusions in a composite. This distribution may be associated with the following statistical parameters and processes within the composite: a spread in shapes of individual inclusions comprising the composite; possible agglomeration of inclusions to clusters; and the effects of multiple scattering and inhomogeneous fields excited by neighboring inclusions. Again, the spectral function is the same for all susceptibilities of a particular composite.

The sum rules,

$$\int_0^1 b(n) \, dn = 1 \quad \text{and} \quad \int_0^1 n b(n) \, dn = \frac{(1-p)}{D} \tag{6}$$

relate the spectral function $b(n)$ to the volume fraction of inclusions p for a macroscopically isotropic composite in D dimensions. The practically important cases are D=3 (an isotropic 3D composite with non-aligned randomly distributed inclusions, the shape of which is arbitrary in the general case) andD=2 (an assembly of infinitely long cylinders). The sum rules provide an agreement of the spectral theory with the LLL mixing rule at $\chi_{\text{incl}} \to 0$.

The spectral theory provides a complete characterization of the frequency dependence of the effective material parameters. The concentration dependence of effective material parameters is implicit in the spectral theory, with the volume fraction involved in the spectral function as a parameter. The analysis of concentration dependences is a powerful tool for understanding properties of composites. However, application of the spectral function approach is not convenient for such analysis.

Another reason that prevents the BM theory from the wide use for the analysis of measured data is that the theory exploits an unknown function, which is difficult to find from the experiment. There are just a few published examples of how to apply the BM theory to the measured data analysis and predicting frequency characteristics of composites [15]. A conventional approach is to accept a functional dependence b(n) as a function of some parameters and to search for these parameters from the measured data [16−18].

Figure 1 shows the calculated spectral functions b(n) for some mixing rules. The spectral function for the MG mixing rule is a delta-function, as is shown in Fig. 1a. The spectral function for the EMT mixing rule presented in Fig. 1b is a semi-circle when plotted as nb(n) against n. Plots d, e, and f show the spectral functions for the McLachlan, Sheng, and Musal–Hahn mixing rules, correspondingly, which are discussed in Subsection 1.4. The latter two plots are composed of several distinct peaks of spectral function even at n=1/3. Other examples of calculated spectral functions for mixing rules are found in [19]. In case of elongated inclusions, or if a composite is composed of inclusions with significantly different aspect ratios, the spectral function may consist of two or larger number of separated peaks. Also, several distinct peaks of the spectral function are found to appear due to the interaction between inclusions in periodical composite structures [8].

The Percolation Theory

A different approach is provided by the percolation theory, see, e.g., [17]. The percolation theory considers the quasi-static permittivity of a metal-dielectric mixture at concentrations close to the percolation threshold. The main assumption of the theory is that the properties of the material are due to statistical properties of large conductive clusters in this case, rather than due to individual properties of inclusions. The theory predicts a power dependence of static permittivity of the mixture on the difference between the concentration, p, and the percolation threshold, p_c:

$$\varepsilon'_{\text{eff}} \propto \left(p_c - p\right)^{-s}, \quad p < p_c, \quad s \approx 0.7$$

$$\varepsilon''_{\text{eff}} \propto \left(p - p_c\right)^{t}, \quad p > p_c, \quad t \approx 1.8$$

$$(7)$$

The values of critical indices, s and t, are believed to be universal, i.e., independent of detailed structure of composite.

A consequence of Eqs. (7) is a power dependence of the real and imaginary permittivity on frequency f:

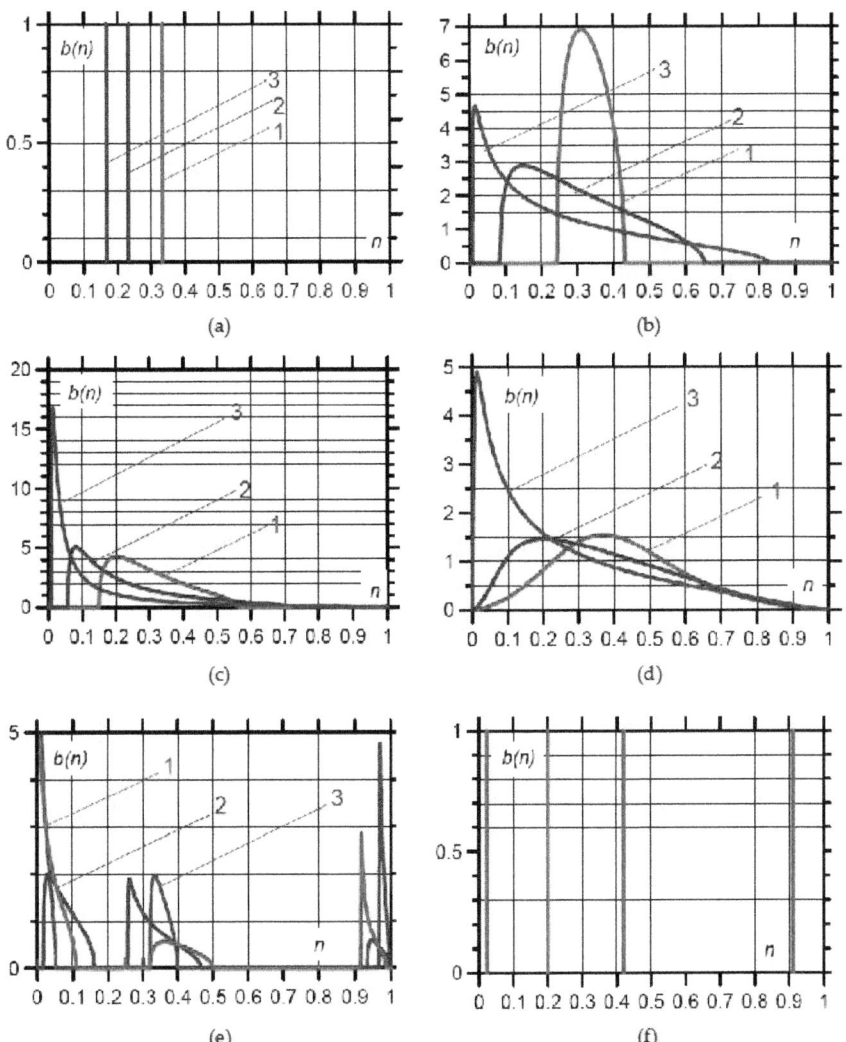

Figure 1: The spectral function, b(x), calculated for various mixing rules: (a) MG, n=1/3: 1 – p=0.1, 2 – p= 0.2, 3 – p=0.5; (b) EMT, n=1/3: p=0.01, 2 – p= 0.1, 3 – p=0.25; (c) Asymmetric Bruggemann's mixing rule: 1 – p=0.1, 2 – p= 0.3, 3 – p=0.6, (d) McLachlan's theory, n=1/3, s=1.8, t=0.7:p=0.01, 2 – p= 0.1, 3 – p=0.25; (e) Sheng theory, n=1/3: 1 – p=0.25, F=0.25,, 2 – p=0.1, F=0.25, 3 – p=0.25, F=0.75; (f) Musal–Hahn theory: n=1/3, p=0.4, F =0.2.

$$\varepsilon'_{\text{eff}} - \varepsilon'_{\text{host}} \propto f^{-\gamma}, \quad \varepsilon''_{\text{eff}} \propto f^{-\gamma} \tag{8}$$

The equality of critical indices for the real and imaginary permittivity established by Eq. (8) follows from the Kramers–Krönig relations, if the frequency dependence of permittivity is governed by law (8) within the whole frequency range. In this case, the dielectric loss tangent, $\varepsilon''/\varepsilon'$, is independent of frequency and equal to $\tan(\pi Y/2)$. It follows from the percolation theory that $Y=s/(s+t)\approx 0.28$. In practice, observed values of Y are typically closer to zero [20].

The physical reason for the powder dependence of the permeability on frequency may be understood as follows. Assume that the frequency response of an individual inclusion in the composite is governed by the Debye frequency dispersion law,

$$\varepsilon(f) = \varepsilon_\infty + \frac{\varepsilon_0 - \varepsilon_\infty}{1 + if\tau},$$

(9)

where ε_0 is the static permittivity, ε_∞ is the optical permittivity, and τ is the characteristic relaxation time, which is reciprocal to the linear relaxation frequency f_{rel}: $\tau = 1/f_{rel}$. The Debye dispersion law governs the frequency dependence of composites filled with conducting inclusions in most cases. When individual inclusions form large clusters of various sizes, a spread of the characteristic relaxation times τ appears. In this case, the total permittivity is written as:

$$\varepsilon(\omega) = \varepsilon_\infty + \int_0^\infty \frac{B(y)\,dy}{1 + i\omega y}$$

(10)

where B(y) is the distribution function of the relaxation times, and

$$y = \tau/\tau_0 \quad \text{and} \quad \int_0^\infty B(y)\,dy = 1$$

(11)

The cumulative dispersion curve becomes more gently sloping. With a special form of the distribution,

$$B(y) = \frac{y}{2\pi} \frac{\sin \beta\pi}{\cosh(1-\beta)y - \cos \beta\pi}$$

(12)

the Cole-Davidson frequency dispersion, see, e.g., [21], is obtained,

$$\varepsilon(f) = \varepsilon_\infty + \frac{\varepsilon_0 - \varepsilon_\infty}{1 + (if\tau)^{1-\alpha}}$$

(13)

Dispersion law (13) involves a frequency region governed by a power frequency dependence of the permittivity. The form of the distribution does not significantly affect the result provided that the distribution is wide, which can be a kind of justification for the percolation theory.

When the property under consideration is the permeability, the percolation behavior is not readily observed [22].

Complex Mixing RuleS

There are many practical scenarios when none of the simple mixing laws agree with the measured data on a practical composite. A classical example is related to carbonyl iron composites. Despite almost perfect spherical shape of carbonyl iron particles, the form factor restored from the volume fraction dependence of the permittivity or permeability frequently differs greatly from 1/3 [18, 23]. The reason is an agglomeration of the inclusions.

Another example is the percolation threshold study in composites composed of the same carbon black and different polymer host matrices [24]. Depending on a polymer, the percolation threshold may vary from 5 to 50%. Polymerization with different polymers results in different morphology of the composites. The reason is agglomeration or de-agglomeration of inclusions, which depends on the properties of the interface between inclusions and the host matrix. The importance of spatial distribution of inclusions in a composite for validity of mixing rules is discussed in [25].

Practically, in describing properties of composites, many other factors must be accounted for. Among these factors, there are the distribution of inclusions in shape [26–28] and size [29]; the presence of an oxide layer on the surface of conducting particles, statistical spread of intrinsic material parameters of inclusions, e.g., their conductivities [30], as well as possible cones of orientations, if elongated particles are aligned or randomly oriented [31]. For these reasons, fitting parameters are typically unavoidable in accurate description of material properties of composites.

Therefore, taking into account peculiarities of a composite morphology may be crucial for accurate description of composite performance, especially in the case of permittivity of metal-dielectric mixtures, where the intrinsic permittivity of inclusions is infinity in the quasi-static case, and the effective permittivity of composite is determined solely by the shape of inclusions.

Conventionally, a composite morphology is accounted for using more complex mixing rules, which involve some fitting parameters. Three examples of such mixing rules are discussed below. These theories combine the above mentioned basic approaches, and allow for introducing appropriate fitting parameters.

A well known example of such combination is the Lichtenecker mixing rule [32], which is written for the case of the effective permittivity as

$$\varepsilon_{\text{eff}}^{k} = p\varepsilon_{\text{incl}}^{k} + (1-p)\varepsilon_{\text{host}}^{k} \tag{14}$$

In Eq. (14), k has a physical meaning of a critical exponent, which is conventionally treated as a fitting parameter to obtain an agreement with measurements. Equation (14) may be considered as an empirical combination of the LLL mixing rule and the percolation theory.

A combination of the EMT mixing rule and the percolation theory is MsLachlan's Generalized Effective Medium Theory [33]:

$$p\frac{\varepsilon_{\text{incl}}^{1/t} - \varepsilon_{\text{eff}}^{1/t}}{\varepsilon_{\text{eff}}^{1/t} + n\left(\varepsilon_{\text{incl}}^{1/t} - \varepsilon_{\text{eff}}^{1/t}\right)} + (1-p)\frac{\varepsilon_{\text{host}}^{1/s} - \varepsilon_{\text{eff}}^{1/s}}{\varepsilon_{\text{eff}}^{1/s} + n\left(\varepsilon_{\text{host}}^{1/s} - \varepsilon_{\text{eff}}^{1/s}\right)} = 0 \tag{15}$$

In this equation, which is also written for the case of permittivity, the EMT equation (2) is supplemented with percolation indices s and t. These indices, together with the effective depolarization factor of inclusions n, are also treated as the fitting parameters.

Another approach for developing complex mixing rules is to divide inclusions in composite into two groups (e.g., a part of inclusions are considered as isolated and the other part are assumed to compose dense clusters [34], or any other way of subdivision into groups), and then to mix these groups with different mixing rules. The value of F, $0<F<1$, a fraction of inclusions attributed to one of the groups, provides a fitting parameter. An example of this approach is Sheng's theory [35]:

$$FP_1 + (1-F)P_2 = 0, \tag{16}$$

where P_1 and P_2 are the effective polarizabilities of the two groups of particles; for the case of spherical inclusions, P_1 and P_2 are found in Sheng's theory as

$$P_1 = \frac{\left(\varepsilon_{\text{eff}} - \varepsilon_{\text{host}}\right)\left(\varepsilon_{\text{incl}} + 2\varepsilon_{\text{host}}\right) + \left(\varepsilon_{\text{host}} - \varepsilon_{\text{incl}}\right)\left(\varepsilon_{\text{eff}} + 2\varepsilon_{\text{host}}\right)p}{\left(2\varepsilon_{\text{eff}} + \varepsilon_{\text{host}}\right)\left(\varepsilon_{\text{incl}} + 2\varepsilon_{\text{host}}\right) + 2\left(\varepsilon_{\text{eff}} - \varepsilon_{\text{host}}\right)\left(\varepsilon_{\text{host}} - \varepsilon_{\text{incl}}\right)p}, \tag{17}$$

$$P_2 = \frac{\left(\varepsilon_{\text{eff}} - \varepsilon_{\text{incl}}\right)\left(2\varepsilon_{\text{incl}} + \varepsilon_{\text{host}}\right) + \left(\varepsilon_{\text{incl}} - \varepsilon_{\text{host}}\right)\left(\varepsilon_{\text{eff}} + 2\varepsilon_{\text{host}}\right)(1-p)}{\left(2\varepsilon_{\text{eff}} + \varepsilon_{\text{incl}}\right)\left(2\varepsilon_{\text{incl}} + \varepsilon_{\text{host}}\right) + 2\left(\varepsilon_{\text{eff}} - \varepsilon_{\text{incl}}\right)\left(\varepsilon_{\text{incl}} - \varepsilon_{\text{host}}\right)(1-p)}. \tag{18}$$

Equation (17) describes the polarizability of a spherical particle consisting of an inclusion material and then coated by a shell of the host matrix material. Equation (18) represents the inverse structure, with the shell made of the inclusion material and the core made of the host matrix material. Both Eqs. (17) and (18) are consistent with the MG formalism. The effective structures are mixed with each other according to the EMT equation (16). An analogous approach is suggested by Musal and Hahn [36], with the only difference that the EMT equation (16) describing a mixture of the two groups is substituted in the MG equation (1). Doyle and Jacobs [34, 37] suggested the model, where the two groups of inclusions comprise isolated inclusions and clusters of closely packed inclusions.

The complex mixing rules are suggested and provide rather good/reasonable agreement with measured data mostly for the concentration dependences of the permittivity in metal-dielectric mixtures. However, these theories may fail when describing frequency dependences of material parameters. The reason is that the complex mixing rules have the spectral function consisting of several isolated peaks even in the case of nearly-spherical inclusions, as is seen in Fig. 1 e and f. A physical meaning can hardly be attributed to these peaks in case of a random composite filled with spherical inclusions, because, as is shown below, the appearance of isolated peaks of spectral function generally results in the appearance of several isolated regions of frequency dispersion of material parameters.

Another approach to the problem of the permittivity dependence on concentration for metal-dielectric mixtures has been suggested by Odelevskiy [38]. He was the first who noticed the analogy between the MG and EMT equations, in which the concentration dependence of the permittivity for conducting inclusions are written as

$$\varepsilon_{eff} = 1 + \frac{1}{n} \frac{p}{1-p}$$
(19)

and

$$\varepsilon_{eff} = 1 + \frac{1}{n} \frac{p}{1-p/n},$$
(20)

respectively. Odelevskiy suggested an equation that generalizes these two theories in the case of a metal-dielectric mixture:

$$\varepsilon_{eff} = 1 + \frac{1}{n} \frac{p}{1-p/p_c}.$$
(21)

In Eq. (21), the form factor n and percolation threshold p_c are the two fitting parameters. With these fitting parameters, the equation demonstrates an excellent agreement with measured data for a variety of different metal-dielectric mixtures [39], if the concentration of inclusions is not very close to the percolation threshold. Equation (21) cannot be considered as an independent mixing rule, because it does not leave a room for the permittivity of inclusions different from infinity.

FREQUENCY-DEPENDENT BEHAVIOR OF COMPOSITES AND VALIDITY OF MIXING RULES

Effective properties of composites in the majority of mixing rules and theories are considered in the quasi-static approximation. Because of this, the frequency dependence of effective material parameter appears due to the difference in frequency dependences of material parameters of constituents.

Frequency dispersion of permittivity in a composite frequently appears due to the different frequency behavior of its dielectric host matrix and of conducting inclusions. Host matrices are typically considered as non-dispersive over a frequency range of interest, while the permittivity of metallic inclusions is imaginary and reciprocal to frequency. There are other dielectric materials possessing dielectric dispersion at microwaves, e.g., water, some ferroelectrics [40], some lossy polymers, but necessity of accounting for this dispersion is a fairly rare.

In contrast, a multitude of magnetic materials exhibit frequency dispersion of permeability at microwaves. The reason is that all magnets lose their magnetic properties at frequencies below several gigahertz, as is shown in Subsection 2.2. These are the microwaves, or even lower frequencies, where the permeability changes from large static permeability to unity. Notice that the intrinsic permeability of magnetic powders is generally unknown. It depends not only on the composition of the material, but also on manufacturing and treatment technology, and the latter dependence may be essential.

In the first-order approximation, the frequency dependence of material parameters may be considered as an assembly of loss peaks accompanied by corresponding frequency dispersion of the real part, according to the Kramers-Kronig relations. In many cases, the Lorentzian (resonance) dispersion law,

$$\chi(f) = \sum_{i=1}^{m} \frac{\chi_{st,i}}{1 + i f/f_{rel,i} - \left(f/f_{res,i}\right)^2},$$

(22)

provides a good fitting of measured dependences of susceptibility χ on frequency f. In Eq. (22), m is the number of the resonance terms involved in the dispersion law, and $\chi_{st,}i$, $f_{rel,}i$, and $f_{res,}i$ are the static susceptibility, relaxation frequency, and resonance frequency attributed to i-th resonance term, respectively.

Frequency-Dependent Behavior of Composites

Almost all mixing rules deduce the effective material parameters from the polarizability,

$$P = \frac{\chi_{incl}}{1 + n \chi_{incl}},$$
$$\tag{23}$$

embedded in either a host matrix or the effective medium. From Eq. (23), two limiting cases are clearly seen, $n\chi_{incl} \ll 1$ and $n\chi_{incl} \gg 1$.

In the case of $n\chi_{incl} \ll 1$, the LLL mixing rule (3) is a rigorous result. For majority of practical cases, Eq. (3) may be rewritten just as the perturbation limit given by Eq. (4). In this case, the effective material parameter is just the intrinsic material parameter multiplied by the volume fraction of inclusions. This means that the effect of interaction between inclusions is negligible. The morphology of the composite, including the shape of inclusions, does not affect the effective material parameter. This case is typical for the microwave permeability of composites filled with either fibrous or platelet inclusions, as well as for all effective susceptibilities at very high frequencies.

In the other limiting case, $n\chi_{incl} \gg 1$, the effective material parameter depends on the morphology only. Here, the effective static susceptibility increases non-linearly with the concentration of inclusions, according to the percolation behavior and Odelevskiy equation (21). It is the case, for which most of the complex mixing rules have been developed. The case is related to the permittivity of metal-dielectric mixtures, since the imaginary part of the permittivity of metal inclusions is so high that the absolute value of the microwave permittivity can be considered as infinite. As to the permeability, this case may be observed in some composites filled with ferromagnetic inclusions of spherical shape, or for low-frequency magnetic materials, whose permeability may be very high.

Let the effective material parameter be considered in a wide frequency range. Assume that the host matrix of the composite is lossless and non-dispersive. Then, the frequency dispersion in the composite is due to the frequency dispersion of inclusions. It is well known that a material parameter of any medium approaches unity with the frequency tending to infinity. Because of that, the case of $n\chi_{incl} \ll 1$ is always observed at very high frequencies, where the LLL mixing rule (3) describes material parameters of composites.

If the intrinsic susceptibility of inclusions is low, the LLL mixing rule is valid for low frequencies as well. In this case, the frequency dependence of any effective material parameter is just proportional to the dependence for the intrinsic material parameter, and the volume fraction of inclusions is the coefficient of proportionality. The loss peak in the composite and the loss peak of inclusions are located at the same frequency. The concentration dependence of the effective parameter is linear over the entire frequency range.

Another possibility is when the inequality $n\chi_{incl} >> 1$ holds at low frequencies. In this case, the frequency dispersion in the composite appears, when the absolute value of $n\chi_{incl}$ is about unity. The loss peak in the composite is shifted towards higher frequencies as compared to the loss peak of inclusions. As the concentration of inclusions increases, the loss peak is shifted to the lower frequencies. At frequencies above the peak, the effective susceptibility is again proportional to the intrinsic susceptibility. At frequencies below the peak, the effective permeability depends mostly on the composite morphology and is independent of the intrinsic susceptibility. The concentration dependence of the effective susceptibility is non-linear.

This case is typical for metal-dielectric mixtures. However, the conductivity of metals is usually too high to provide a loss peak of permittivity at microwaves. The microwave permittivity for most metal-dielectric composites may be considered as non-dispersive and low-loss. An exception is the percolation behavior, which will be discussed in Section 3.

If the frequency dependence of the intrinsic material parameter is Lorentzian (22) with $m=1$, and the mixing rule describing the composite is the MG, then the frequency dependence of the effective material parameter is Lorentzian as well. The parameters of the dispersion law for the effective material parameters are given by the simple equations [41]

$$\chi_{st,eff} = \frac{\chi_{st,incl}\, p}{\chi_{st,incl}\, n(1-p)+1},$$

(24)

$$f_{rel,eff} = f_{rel,incl}\left(\chi_{st,incl}\, n(1-p)+1\right),$$

(25)

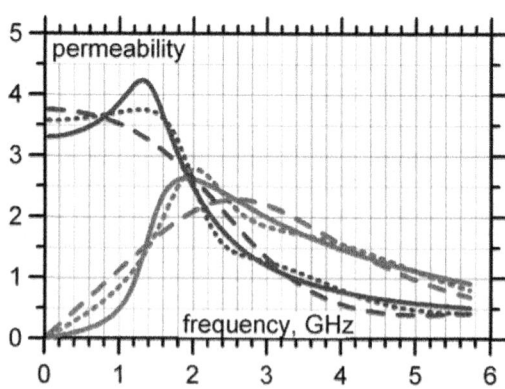

Figure 2: The frequency dependence of effective permeability of a composite calculated by the EMT (2) with $p=0.25$ and $n=1/3$ (solid curves, blue curve for real permeability and red curve for imaginary permeability). Inclusions in the composite exhibit the Lorentzian frequency dispersion (22) with $m=1$, $\chi_{st}=100$, $f_{res}=1$ GHz, and $f_{rel}=2$

GHz. The dashed curves are the best fit of the solid line with the Lorentzian dispersion law with m=1, the dotted curves — with a sum of two Lorentzian terms, m=2.

and

$$f_{res,eff} = f_{res,incl} \sqrt{\chi_{st,incl} \, n(1-p)+1} ,$$

(26)

where subscript "incl" indicates the Lorentzian parameters of the intrinsic permeability of inclusions and subscript "eff" is related to the Lorentzian parameters of the effective susceptibility of composite. It is clearly seen from the equations that the non-linear concentration dependence of static susceptibility is accompanied by a low-frequency shift of both the characteristic frequencies. A general validation of this fact is given in the next subsection.

As is seen from Eqs. (25) and (26), the MG mixing rule retains the shape of effective susceptibility loss peak characteristic for the intrinsic susceptibility of inclusions. From the standpoint of the BM spectral theory, the reason is that additional loss due to mixing may arise over the entire range of effective form factors, where the spectral function has non-zero values. The spectral function for the MG mixing law is a delta-function, therefore, additional loss, which may distort the loss peak, does not appear.

Other mixing rules are characterized by a spectral function of a finite width and may therefore result in distorted shape of the loss peak. Figure 2 shows the frequency dependence of effective permeability of a composite calculated by the EMT (2). Inclusions in the composite are assumed to exhibit the Lorentzian frequency dispersion (22) with m=1. In the figure, the dashed curves are the best fit of the calculated permeability with the Lorentzian dispersion law with m=1, the dotted curves are obtained for the sum of two Lorentzian terms, m=2. It is seen that the EMT produces a large distortion of the Lorentzian dispersion curve, when the concentration is close to the percolation threshold. The distortion has a form of the increased loss at the high-frequency slope of the loss peak, because the spectral function peak for the EMT is extended to the region of large arguments, see Fig. 1b.

Integral Relations for The Frequency Dependences In Composites

The low-frequency shift of the loss peak appearing with increasing volume fraction and accompanied by non-linear concentration dependence of static susceptibility is a general rule. Let us consider two integrals,

$$I_1 = \frac{2}{\pi} \int_0^\infty \chi'' f \, df \quad \text{and} \quad I_2 = \frac{2}{\pi} \int_0^\infty \chi' \, df ,$$

(27)

which are analogous to the well-known sum rule for the Kramers-Kronig relations,

$$\chi_{st}' = \frac{2}{\pi} \int_0^\infty \frac{\chi'' df}{f}.$$

(28)

The difference between (28) and (27) is that the values of I_1 and I_2 are determined by the high-frequency asymptote of the susceptibility, rather than by the low-frequency asymptote, which defines the value of integral (28). In composites, this asymptotic behavior is governed by the LLL mixing law. Therefore, integrals I_1 and I_2 for any composite are equal to the corresponding values for the bulk material of inclusions multiplied by the volume fraction of inclusions [42,43]

$$I_{i,\text{composite}} = p I_{i,\text{inclusions}}.$$

(29)

Consideration of Eq. (29) makes sense if the integrals are convergent and have a non-zero value. For I_1, this is true for the Lorentzian dispersion law (22) that has the high-frequency asymptote given by:

$$\chi(f) \approx -\chi_{st}\left(\frac{f_{res}}{f}\right)^2 + i\chi_{st}\left(\frac{f_{res}}{f}\right)^3 \frac{f_{res}}{f_{rel}}$$

(30)

For I_2, the convergence is provided by the Debye dispersion law (9), which is the limiting case of (22) at $f_{res} \to \infty$ and has the high-frequency asymptote represented as:

$$\chi(f) \approx i\chi_{st}\left(\frac{f_{rel}}{f}\right) + \chi_{st}\left(\frac{f_{rel}}{f}\right)^2.$$

(31)

In the theory of magnetic material, these integrals are employed to validate ultimate values of high-frequency permeability. The corresponding constants for magnetic materials depend on the saturation magnetization of the material, M_s. If the frequency dependence of effective permeability is either single-term Lorentzian or Debye, then the values of the integrals are related to the static magnetic susceptibility and the resonance frequency

$$I_1 = p\kappa(\gamma M_s)^2 \approx \chi_{st,\text{eff}} f_{res,\text{eff}}^2,$$

(32)

$$I_2 = p\kappa(\gamma M_s) \approx \chi_{st,\text{eff}} f_{rel,\text{eff}}.$$

(33)

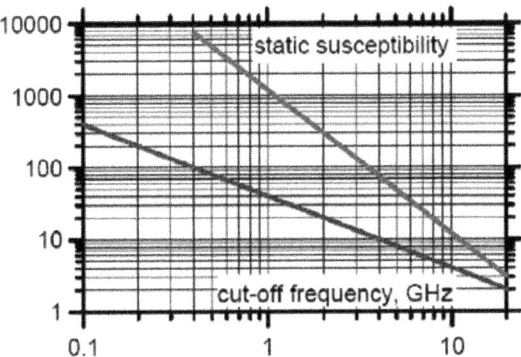

Figure 3: The static susceptibility as a function of cut-off frequency calculated with Acher's law (32) (red line, $\kappa=1/3$) and Snoek's law (33) (blue line, $\kappa=2/3$). In both the cases, $M_s=2.15$ T, p=1.

In Eqs. (32) and (33), $\gamma\approx3$ GHz/kOe is the gyromagnetic ratio, and κ is the randomization factor. For I_2, typically $\kappa=2/3$; for I_1, different possibilities are discussed in [14]. Equations (32) and (33) represent the well known Acher's law [44] and Snoek's law [45], respectively. Then I_1 has a meaning of Acher's constant, and I_2 is Snoek's constant. For most materials, Snoek's law is valid, which involved Debye frequency dependence and integral I_2. For some materials, such as hexagonal ferrites and thin ferromagnetic films, Acher's law is valid, so that integral I_1 is calculated as Eq. (32), and much larger high-frequency permeability values can be obtained. The laws (32) and (33) are used for estimating high-frequency magnetic behavior of materials. A magnetic material may have high permeability value at frequencies below the cut-off frequency, which is the least of f_{res} and f_{rel}, where the permittivity falls to values close to unity. As the saturation magnetization of magnetic materials is typically below approximately 2 T, it follows from (32) and (33) that magnetic materials with high static permeability are permeable at frequencies of microwave range or lower.

Figure 3 shows the ultimate values of the static magnetic susceptibility as a function of the cut-off frequency calculated with Acher's law (32), red line, and Snoek's law (33), blue line. In both cases, $M_s=2.15$ T and p=1, which corresponds to a homogeneous sample of pure iron. For Snoek's law, $\kappa=2/3$; for Acher's law, $\kappa=1/3$ is accepted, which corresponds to random distribution of thin platelets. It is seen from the figure that with low values of the cut-off frequency, below 1 GHz, Acher's law enables a large advantage over Snoek's law in feasible values of the static permeability. At higher frequencies, this advantage eliminates, and both the laws permits rather small ultimate values of static permeability with cut-off frequencies of several dozen gigahertz.

For the permittivity of a metal-dielectric mixture, the frequency dependence is of Debye type, and an analogue of Snoek's law may be introduced. As $\varepsilon''_{incl}=2\sigma/f$, where σ is the conductivity of inclusions, the analogue of Snoek's constant for permittivity would be just the doubled conductivity of inclusions,

$$I_2 = 2p\sigma.$$

(34)

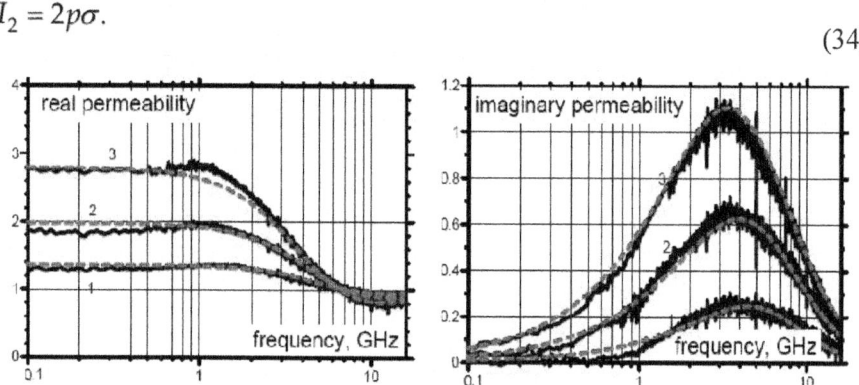

Figure 4: The measured frequency dependence of permeability of hexagonal ferrite composites (black curves, left: the real part, right: the imaginary part) for the Co_2Z composites. The volume fractions of ferrite are: 1, p=0.1; 2, p=0.3; and 3, p=0.5. The red curves show the results of fitting the measured data with the Lorentzian dispersion law (22) with m=1 [41].

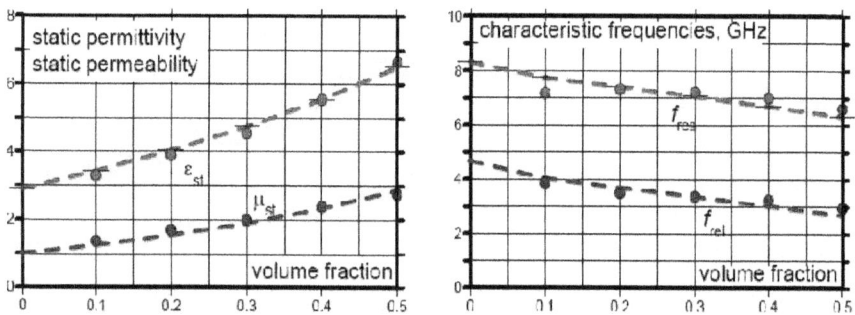

Figure 5: Left: the static permittivity (red dots) and static permeability (blue dots); right: the resonance frequency (red dots) and relaxation frequency (blue dots). The data are obtained for hexagonal ferrite composites by fitting the measured frequency dependences of permeability with the Lorentzian dispersion law (22) with m=1. The curves are the best fit of corresponding dots with Eqs. (24–26) [41].

Applicability of The Mg Mixing Rule

The MG mixing rule usually agrees closely with the measured data, when $n\chi_{incl} \sim 1$. This is a frequent occasion for the microwave permeability of magnetic composites. The intrinsic permeability of magnetic materials does not exceed several units at microwaves due to the fast decrease with frequency, according to Snoek's and Acher's laws. With these relatively low intrinsic permeability values, the dependence of the effective material parameters on the shape of inclusions appears. In particular, a low-frequency shift of the loss peak is observed as p increases. However, the dependence is still weak and may therefore be characterized by an averaged demagnetization factor n.

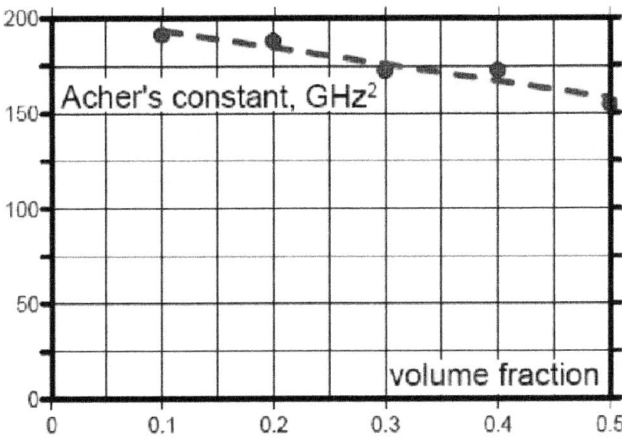

Figure 6: The measured ratio of Acher's constant to the volume fraction plotted against the volume fraction (dots). The line show the linear fit of the measured data [41].

An example of measured data having a good agreement with the MG mixing rule is taken from [41], where composites filled with powders of hexagonal ferrite have been studied. Figure 4 shows the measured microwave permeability for three of the samples. Application of the single-term Lorentzian dispersion law (22) provides a good agreement with the measured data for all volume fractions. This is seen from Fig. 5, where the static permittivity and the Lorentzian characteristic frequencies, obtained by the best fits of the measured magnetic dispersion curves, are plotted as functions of volume fraction. The curves in the figure are obtained by fitting the experimental points (dots) with Eqs. (24–26). For the bulk hexagonal ferrite, the retrieved static values are $\varepsilon_{st,incl}=16$ and $\mu_{st,incl}=11$, and $n\approx0.33$, which indicates that the ferrite particles are of nearly spherical shape. Therefore, $n\chi st,_{incl}$ is the range from 3 to 5, and is reasonably close to unity. The measured data on the microwave material parameters of the composites under study agree with the MG mixing rule calculations, which is evidenced by close agreement of the dots and the fitting curves in Fig. 5.

However, an accurate analysis of the data reveals some disagreement. Acher's constant of the composites, calculated from the data for different volume fractions does not agree with Eq. (32). AsFig. 6 shows, Acher's constant depends on the volume fraction of inclusions, which should not be the case. The reason could be a distribution of shapes of individual inclusions that may result in deviation of the morphology from that postulated in the MG approach. This problem is discussed in more details in the next Subsection.

Account for the Distribution In Shapes Of Inclusions

A case, which may require a sophisticated mixing rule, is a composite filled with conducting ferromagnetic inclusions, whose both permittivity and permeability must be predicted, for example, to describe electromagnetic performance of the composite. In this case, two products of the form factor and the static susceptibility are involved, for the dielectric and magnetic susceptibility, which enlarges the range of variation of this value with a result of necessity for a more sophisticated theory to obtain better agreement between the measured data and theory.

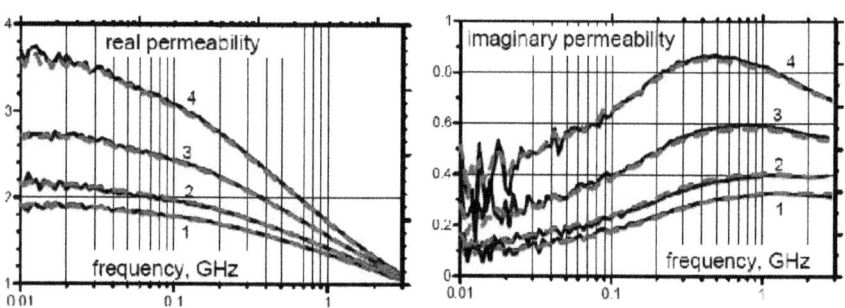

Figure 7: Black curves: the measured frequency dependencies of permeability of composites filled with milled iron powder (black lines), left − real permeability, right − imaginary permeability. Red curves − fitting of the measured data with theory (35). Volume fractions of inclusions are 15.0% (1), 17.7% (2), 23.6% (3), 30.3% (4) [39].

Recently, a new theory, which unites the MG and EMT approaches, has been proposed [46]. This theory allows for introducing the percolation threshold through a general quadratic equation, the same as the EMT, postulating two requirements to the solution. The On the one hand, the solution must be consistent with the LLL mixing rule (3) for the case of low intrinsic material parameter; on the other hand, it should satisfy the Odelevskiy equation (21) for the case of intrinsic material parameter tending to infinity. This produces a unique solution for the equation, which can be considered as a new mixing rule, which generalizes the EMT and MG mixing rules,

$$p\frac{\chi_{\text{eff}}}{\chi_{\text{incl}}} + \frac{1-p}{Dpn}\frac{1}{1/p - 1/p_c - 1/n/\chi_{\text{eff}}} = 1,$$ (35)

where D is the dimensionality of composite. Mixing rule (35) involves two fitting parameters: the effective form factor of inclusions, n, and the percolation threshold, p_c, that can be found from the concentration dependence of the effective permittivity. In fact, these parameters are related to peculiarities of morphology of composites, such as the distribution of inclusions in shape.

The derivation of Eq. (35) is based on the assumption that the spectral function has a single wide peak. It is shown [46] that Eq. (35) allows for a variation of these parameters over the ranges,

$$\sqrt{4/D} - 1 < n < 1/D,$$

$$\frac{1}{2}(Dn(1+n) - n\sqrt{D}\sqrt{D(1+n)^2 - 4}) < p_c < 1.$$ (36)

These conditions correspond to the case of nearly-spherical inclusions. Derivation of similar approach for composites filled with highly-elongated inclusions, such as thin platelets or fibers, must incorporate a spectral function comprising two separated peaks, which would require more sophisticated mathematical approaches. However, to develop such approach is not a challenging problem. As is shown above, the dilute limit approximation is sufficient for the analysis of microwave magnetic performance of such composites.

With the fitting parameters retrieved from the concentration dependence of permittivity, the intrinsic permeability of inclusions may be found from the measured effective permeability at each volume fraction of inclusions in the composite, as is described in [39]. An agreement of the data on the intrinsic permeability of inclusions found from different concentrations of inclusions provides an additional test for the validity of the mixing rule. It is found that the theoretical predictions agree closely with the measured microwave permittivity and permeability of composites filled with milled Fe powders [39], see Fig. 7. In the figure, the intrinsic permeability of inclusions was calculated for each concentration of inclusions, after which the average value was used to calculate the theoretical curve for each concentration. This is the reason for the noise observed in the theoretical curves in Fig. 7.

COMPOSITES WITH FIBROUS INCLUSIONS

Measured Microwave Permittivity of Fiber-Filled Composites

Frequency dispersion of permittivity typically is not observed in composites over the microwave range. One of the rare examples of microwave dielectric dispersion is provided by composites filled with carbonized organic fibers. The conductivity of such fibers is much lower that that of metals. The thickness of the fibers is about a few microns, and their length can be on the order of several millimeters. The form factor of the fiber is very low, and the region of frequency dispersion may be at microwaves, as is seen in Fig. 8 [47].

Figure 8 shows the measured frequency dependence of permittivity for a composite filled with carbon fibers with length l=1.5 mm, thickness d=8 μm, and resistivity of 10 000 Ohm×cm. The volume fraction of the fibers in the composite is p=0.01%. The sample is a sheet polymer-based composite of thickness of less than 1 mm. Fibers are parallel to the sheet material plane, and they are distributed and oriented randomly in this plane. Experimental details are given in [47]. The frequency dependence of permittivity is of the Debye type. The low-frequency permittivity varies linearly with the volume fraction. The measured frequency and concentration dependences of permittivity agree well with the dilute limit approximation (4), written for the case under consideration as

$$\varepsilon_{\text{eff}} = \varepsilon_{\text{host}} \left(1 + p\kappa \frac{2i\sigma/f - \varepsilon_{\text{host}}}{\varepsilon_{\text{host}} + n\left(2i\sigma/f - \varepsilon_{\text{host}}\right)} \right),$$

$$(37)$$

where σ is the conductivity of the fibers, κ is a factor describing the averaged polarizability of inclusions, and n is the depolarization factor of the fibers,

$$n = \frac{d^2}{l^2} \ln\frac{l}{d}.$$

$$(38)$$

The value of κ =1/3 in Eq. (37) for the case under consideration, that is a product of the value of ½, which accounts for the isotropic in-plane orientations of the fibers in a sheet sample, and the value of 2/3, which accounts for cylindrical shape of fibers instead of ellipsoidal shape considered by the theories.

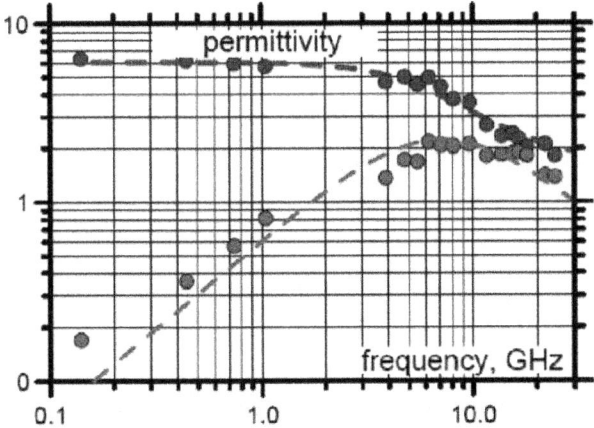

Figure 8: The measured frequency dependence of real (blue dots) and imaginary (red dots) permeability of a composite filled with carbon fibers of 1.5 mm in length with the resistivity of 10 000 Ohm×cm. The volume fraction of the fibers is 0.01% [47]. Curves are the result of fitting of measured data with the Debye dispersion law.

The type of the frequency dependence observed in fiber-filled composites is determined by the conductivity of fibers. In the general case, the dielectric dispersion curve is of the Lorentzian type with the parameters written as [48]

$$\chi_{st} = \frac{1}{3}\frac{\varepsilon_{host}p}{\ln l/d}\left(\frac{l}{d}\right)^2 , \ f_{rel} \approx \frac{2\sigma}{\varepsilon_{host}}\left(\frac{d}{l}\right)^2 \ln l/d , \ f_{res} \approx \frac{c}{2l\sqrt{\varepsilon_{host}}}$$

(39)

The resonance of the permittivity arises from the half-wavelength resonance excited within the fibers.

Figure 9 shows the measured frequency dependence of permittivity for a composite filled with aluminum-coated fibers of 10 mm long [47]. The volume fraction the fibers is 0.01%. Due to high conductivity of the fibers, the frequency dependence of permittivity is of pronounced resonance (Lorentzian) type.

It is seen from Fig. 9 that the quality factor of the dielectric resonance is much lower than that predicted by Eq. (39). This is because Eq. (39) does not account for the radiation resistance of the fibers. The radiation resistance of a half-wavelength dipole is approximately 75 Ohm in the free space, which is much larger than the ohmic resistance of the fiber, and contributes dominantly to the quality factor of the resonance.

In fact, such composites behave as a kind of a metamaterial over the frequency range near the resonance, because they contain inhomogeneities, whose characteristic dimensions are close to the wavelength, and the principal features of their dielectric dispersion depend on the resonance scattering on the fibers. This is also evidenced by the facts that the measured permittivity is less than that produced by the MG mixing law, and that the radiation resistance makes a dominant contribution into the quality factor of the dielectric resonance. Rigorously, metamaterials cannot be described in terms of effective material parameters. However, an experimental observation of deviation of microwave performance of the composites from Fresnel law has required special measurement conditions, see [49] for details.

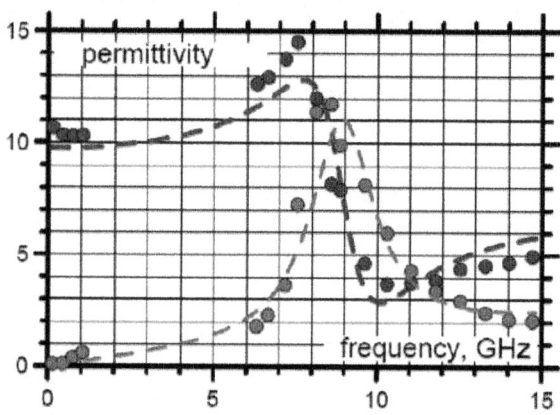

Figure 9: The measured frequency dependence of real (blue dots) and imaginary (red dots) permeability of a composite filled with aluminum-coated fibers 20 mm long. The volume fraction of the fibers is 0.01% [47]. Curves are the result of fitting of measured data with the Lorentzian dispersion law.

Theories for the Effective Properties of Fiber-Filled Composites

A typical feature of fiber-filled composites is a low value of the percolation threshold: $p_c \propto d/l$, see [47] for the measured data. Although the percolation threshold is conventionally considered as a structure-dependent parameter, the dependence has been validated with composites based on a random mixture of conducting and non-conducting fibers, so that the dependence on agglomeration would be minimized. The standard EMT produces even lower values, $p_c \propto (d/l)^2$, and a large disagreement with the measured permittivity values at concentrations close to the percolation threshold can be observed. Mixing rules for the fiber-filled composites were primarily aimed at obtaining proper dependence of the percolation threshold on the aspect ratio of fibers.

Historically, the first theories describing the effective properties of fiber-filled composites have been suggested in [37] and [50]. However, the theory [37] describes the case of infinite conductivity of inclusions and is not suitable for describing of frequency dependences in metal-dielectric composites. Theory [50] is a modification of the EMT. It is based on the assumption of strong anisotropy of the effective medium in the vicinity of a particular fiber, which results in the equations

$$\frac{p}{3}\left(\frac{\varepsilon_{\text{eff,II}}}{\varepsilon_{\text{eff,II}}+n\left(\varepsilon_{\text{incl}}-\varepsilon_{\text{eff,II}}\right)}+\frac{2\varepsilon_{\text{eff},\perp}}{\varepsilon_{\perp}+n_{\perp}\varepsilon_{\text{incl}}\left(\varepsilon_{m}-\varepsilon_{\text{eff},\perp}\right)}\right)+\frac{\left(1-p\right)\varepsilon_{\text{eff},\perp}}{\varepsilon_{\perp}+n\left(\varepsilon_{\text{host}}-\varepsilon_{\text{eff},\perp}\right)}=1,$$

(40)

$$\frac{p}{3}\left(\frac{\left(\varepsilon_{\text{eff,II}}-\varepsilon_{\text{incl}}\right)}{\varepsilon_{\text{eff,II}}+n\left(\varepsilon_{\text{incl}}-\varepsilon_{\text{eff,II}}\right)}+\frac{2\left(\varepsilon_{\text{eff},\perp}-\varepsilon_{\text{incl}}\right)}{\varepsilon_{\perp}+n_{\perp}\left(\varepsilon_{\text{incl}}-\varepsilon_{\text{eff},\perp}\right)}\right)+\frac{\left(1-p\right)\left(\varepsilon_{\text{eff},\perp}-\varepsilon_{\text{host}}\right)}{\varepsilon_{\perp}+n\left(\varepsilon_{\text{host}}-\varepsilon_{\text{eff},\perp}\right)}=0.$$

(41)

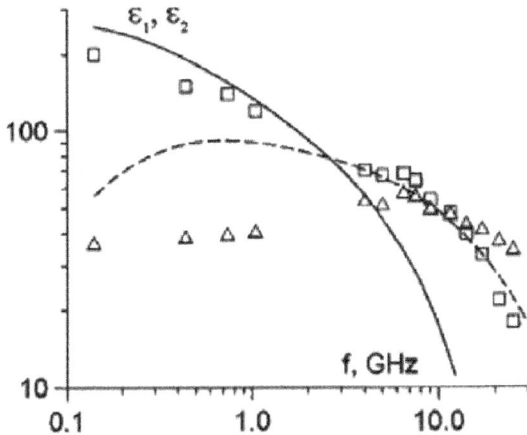

Figure 10: Dots: the measured dielectric dispersion curves for a composite filled with carbon fibers 1.8 mm long with the resistivity of 1400 Ohm×cm. The volume concentration of the fibers is 0.52%. Curves: calculation by Eq. (42) [47].

The two equations (40) and (41) are used for the search of two values of the effective permittivity, $\varepsilon_{\text{eff,II}}$ and $\varepsilon_{\text{eff},\perp}$, in the directions parallel and perpendicular to a fiber, respectively; n is given by Eq. (38), $n_{\perp}=(1-n)/2$, and the observed effective permittivity of the composite is found by averaging of $\varepsilon_{\text{eff,II}}$ and $\varepsilon_{\text{eff},\perp}$. The value of the randomization factor κ, defined in the Subsection 2.1, equals to 1/3, and is substituted in Eqs. (40) and (41). The theory [50] predicts correctly the dependence of the percolation threshold on the aspect ratio of fibers, but disagrees with the dilute limit approximation, and, therefore, with the measured permittivity of composites at low concentrations of fibers.

Theory [51] allows for a better quantitative agreement over a wide range of volume fractions below the percolation threshold. The theory is based on the assumption that, in the vicinity of a particular fiber, the permittivity of effective medium ε^* is a function of the distance from to fiber z:

$$\varepsilon^*(z) = \varepsilon_{host} + \left(\varepsilon_{eff} - \varepsilon_{host}\right)\left(z/ax\right) \quad \text{at} \quad z < ax$$

$$\varepsilon^*(z) = \varepsilon_{host} \quad \text{at} \quad z \geq ax \tag{42}$$

where x is a parameter of the theory. This assumption results in the EMT equation written as

$$\frac{\varepsilon_{incl}}{3\varepsilon_{eff}} \frac{p}{1 + \left(d^2 \varepsilon_{incl}/l^2 \varepsilon_{host}\right)\ln\left(1 + \left(l x \varepsilon_{host}\right)/\left(d\varepsilon_{eff}\right)\right)} + 3\frac{\varepsilon_{host} - \varepsilon_{eff}}{\varepsilon_{host} + 2\varepsilon_{eff}} = 0, \tag{43}$$

which is conventionally used in microwave studies of fiber-filled composites, see, e.g., [52].

There is lack of measured data on the frequency-dependent dielectric performance of fiber-filled composites near the percolation threshold in the literature; one of examples of the data is given in Fig. 10 [45]. It is seen from the figure that the EMT predicts a gradual shift of the loss peak. The measured dielectric loss peak differs from that predicted by the theory. In contrast, the measured variation of the loss peak appearing as approaching the percolation threshold looks like a rise of the low-frequency loss level, with a well-defined trace of the loss peak associated with individual fibers.

This difference between the theory and measurements may be associated not with the geometrical distribution of shapes of conducting clusters, as the percolation theory suggests, but with other low-frequency loss mechanisms near the percolation threshold. For example, imperfect electric contacts between fibers comprising a conductive cluster may contribute to the low-frequency loss [20]. The conductivity of such contacts must be much lower than the conductivity of the fibers. Therefore, imperfect contacts may result in a large low-frequency shift of dielectric loss. Because of a low value and wide distribution of the conductivity of contacts, this loss forms a very smooth dispersion curve, which is seen in Fig. 10.

The same may be true for composites filled with carbon black or carbon nanotubes, which are known to have the percolation type of frequency dispersion at microwaves, see, e.g., [20]. Dielectric loss appearing at low frequencies might be associated with very prolate conductive clusters, if the conductivity of clusters is on the same order of magnitude as the conductivity of inclusions. Account for the imperfect contacts would allow for more realistic assumptions on the shape of conducting clusters.

In principle, the effect of contacts may be understood as a presence of a comparatively low-conductive shell covering the surface of conducting inclusions. For an individual inclusion, the presence of such a shell leads to a low-frequency shift of the loss peak, without change in its shape. To get an agreement with the measured data, a distribution of these conductivities should be included in the model. By the analogy to the Cole-Cole dispersion law, such distribution would results in the power frequency dependences of the permittivity. Hence the difference between measured critical indices and universal values derived from geometrical considerations can be observed, but there is no theory explaining and quantifying such phenomena in the literature.

The available data of the microwave permeability of composites filled with magnetic fibers are consistent with the dilute limit approximation [53, 54].

CONCLUSIONS

The problem of describing of the effective permittivity as a function of concentration of inclusions in a metal-dielectric mixture is well studied. However, newly developed mixing rules still appear in the literature. This means that the solution for the problem is not satisfactory to some extent, and is typically related to the description of frequency dependences of material parameters.

The above consideration allows for determining the validity limits of various mixing rules. These limits are dependent on the difference between the susceptibilities of inclusions and the host matrix, and on the elongation of inclusions.

For microwave permeability, the difference is typically not high, and the effective properties of composites are well described by the MG mixing rule. For lower frequencies, the intrinsic permeability may be high, and a more sophisticated mixing rule may be needed. For composites containing platelet and fibrous magnetic inclusions, the microwave permeability is described by the dilute limit approximation. The same is true for composites with dielectric fibers.

For microwave permittivity of a metal-dielectric mixture, the difference is typically large, and the effective properties are determined by the morphology of the composite. But fitting of measured data to the theoretical results is typically rather simple, because the frequency dispersion of permittivity is a rare occasion at microwaves. In metal-dielectric composites, the region of frequency dispersion is located at much higher frequencies, as can be estimated from typical conductivity of metals and feasible dimensions of inclusions.

For simultaneous modeling of the permittivity and permeability of composites with conducting inclusions, sophisticated mixing rules are unavoidable, with an account for a distribution of inclusions in shape. This case is the most difficult, because both concentration and frequency dependences of material parameters may be non-trivial.

ACKNOWLEDGEMENT

K. Rozanov acknowledges the partial financial support of the work from the RFBR, grants no. 12-02-91667 and 12-08-00954. M. Koledintseva acknowledges the partial support by the U.S. NSF Grant No. 0855878. The authors also thank Alexei Koledintsev for his assistance and valuable comments regarding technical English writing.

REFERENCES

1. Koledintseva MY, Rozanov KN, Drewniak JL2011Engineering, modeling and testing of composite absorbing materials for EMC applications, In: Adv. in Composite Materials – Ecodesign and Analysis, ed. B. Attaf, InTech, 978-9-53307-150-3Ch. 13, 291

2. Garnett JCM1904Colours in Metal Glasses and in Metallic Films. Phil. Trans. R. Soc. Lond. 203: 385–420.

3. Bruggeman DAG1935Berechnung Verschiedener Physikalischer Konstanten von Heterogenen Substanzen. Ann. Phys. (Leipzig) 24: 636–679.

4. Landau LD, Lifshitz EM1984Electrodynamics of Continuous Media, Pergamon, 474 p.

5. H. Looyenga, 1965Dielectric Constants of Heterogeneous Mixtures. Physica. 31: 401–406.

6. M. Born, E. Wolf, 1986Principles of Optics. 6 Ed., Pergamon, 854 p.

7. D. Polder, J. H. van Santen, 1946The Effective Permeability of Mixtures of Solids. Physica. 12: 257–271.

8. D. J. Bergman, D. Stroud, 1992Physical Properties of Macroscopically Inhomogeneous Media. Solid. State Phys. 46: 147–269.

9. Z. Hashin, S. Shtrikman, 1962A Variational Approach to the Theory of the Effective Magnetic Permeability of Multiphase Materials J. Appl. Phys., 33: 3125–3131.

10. B. Sareni, L. Krahenbuhl, A. Beroual, C. Brosseau, 1996Effective Dielectric Constant of Periodic Composite Materials. J. Appl. Phys. 80: 1688–1696.

11. Sihvola AH1999Electromagnetic Mixing Rules and Applications. IET, 284 p.

12. S. Torquato, S. Hyun, 2001Effective Medium Approximation for Composite Media: Realizable Single-Scale Dispersions. J. Appl. Phys. 8917251729

13. Reynolds JA, Hough JM1957Formulae for Dielectric Constant of Mixtures. Proc. Phys. Soc. B 70: 769−775.

14. Lagarkov AN, Rozanov KN2009High-Frequency Behavior of Magnetic Composites. J. Magn. Magn. Mater. 32120822092

15. Day AR, Grant AR, Sievers AJ, Thorpe MF2000Spectral Function of Composites from Reflectivity Measurements. Phys. Rev. Lett. 84: 1978−1981.

16. W. Theiss, 1996The Dielectric Function of Porous Silicon- How to Obtain It and How to Use It. Thin Solid Films. 276712

17. K. Ghosh, R. Fuchs, 1988Spectral Theory for Two-Component Porous Media. Phys. Rev. B: 385222

18. Osipov AV, Rozanov KN, Simonov NA, Starostenko SN2002Reconstruction of Intrinsic Parameters of a Composite from the Measured Frequency Dependence of Permeability. J. Phys.: Condens. Matter 1495079523

19. Goncharenko AV, Lozovski VZ, Venger EF2000Lichtenecker's Equation: Applicability and Limitations. Optics. Commun. 174: 19−32.

20. L. Liu, S. Matitsine, Y. B. Gan, L. F. Chen, L. B. Kong, K. N. Rozanov, 2007Frequency Dependence of Effective Permittivity of Carbon Nanotube Composites. J. Appl. Phys. 101: 094106.

21. Jonscher AK1983Dielectric Relaxation in Solids, Chelsea Dielectrics Press, 1983, 380 p.

22. J. L. Mattei, M. Le Floc'h, 2003Percolative Behaviour and Demagnetizing Effects in Disordered Heterostructures. J. Magn. Magn. Mater. 257335345

23. K. C. Pitman, M. W. Lindley, D. Simkin, J. F. Cooper, 1991Radar Absorbers: Better by Design. IEE Proc. F- Radar and Signal Processing 138223228

24. K. Miyasaka, K. Watanabe, E. Jojima, H. Aida, M. Sumita, K. Ishikawa, 1982Electrical Conductivity of Carbon-Polymer Composites as a Function of Carbon Content. J. Mater. Sci. 1716101616

25. H. L. Duan, B. L. Karihaloo, J. Wang, X. Yi, 2006Effective Conductivities of Heterogeneous Media Containing Multiple Inclusions with Various Spatial Distributions. Phys. Rev. B 73: 174203.

26. L. Gao, J. Z. Gu, 2002Effective Dielectric Constant of a Two-Component Material with Shape Distribution. J Phys. D- Appl. Phys. 35267271

27. Goncharenko AV2003Generalizations of the Bruggeman Equation and a Concept of Shape-Distributed Particle Composites. Phys. Rev. E 68: 041108.

28. Koledintseva MY, Chandra SKR, DuBroff RE, Schwartz RW2006Modeling of Dielectric Mixtures Containing Conducting Inclusions with Statistically Distributed Aspect Ratio. PIER 66213228

29. A. Spanoudaki, R. Pelster, 2001Effective Dielectric Properties of Composite Materials: The Dependence on the Particle Size Distribution. Phys. Rev. B 64: 064205.

30. Koledintseva MY, DuBroff RE, Schwartz RW, Drewniak JL2007Double Statistical Distribution of Conductivity and Aspect Ratio of Inclusions in Dielectric Mixtures at Microwave Frequencies. PIER 77193214

31. Koledintseva MY, DuBroff RE, Schwartz RW2009Maxwell Garnett Rule for Dielectric Mixtures with Statistically Distributed Orientations of Inclusions. PIER 99 131−148.

32. K. Lichtenecker, 1926Die Dielektrizitätskonstante Natürlicher und Künstlicher Mischkörper. Physikal. Z. 27115158

33. Priou. A. Mc Lachlan, I. Chenerie, E. Isaak, F. Henry, 1992Modeling the Permittivity of Composite Materials with a General Effective Medium Equation. J. Electromagn. Waves Appl. 610991131

34. Doyle WT, Jacobs IS1990Effective Cluster Model of Dielectric Enhancement in Metal-Insulator Composites. Phys. Rev. B 4293199327

35. P. Sheng, 1980Theory for the Dielectric Function of Granular Composite Media. Phys. Rev. Lett. 45: 60−63.

36. Musal HM, Hahn HT, Bush GG1988Validation of Mixture Equations for Dielectric-Magnetic Composites. J. Appl. Phys. 6337683770

37. Doyle WT, Jacobs IS1992The Influence of Particle Shape on Dielectric Enhancement in Metal-Insulator Composites. J. Appl. Phys. 7139263936

38. Odelevskiy VI1947The Calculation of Generalized Conductivity of Heterogeneous Systems. Ph. D. Thes., Moscow, 110 p.

39. Rozanov KN, Osipov AV, Petrov DA, Starostenko SN, Yelsukov EP2009The Effect of Shape Distribution of Inclusions on the Frequency Dependence of Permeability in Composites. J. Magn. Magn. Mater. 321738741

40. M. Li, A. Feteira, D. C. Sinclair, A. R. West, 2007Incipient Ferroelectricity and Microwave Dielectric Resonance Properties of CaCu2.85Mn0.15Ti4O12 Ceramics. Appl. Phys. Lett. 91: 132911.

41. Rozanov KN, Li ZW, Chen LF, Koledintseva MY2005Microwave Permeability of Co2Z Composites. J. Appl. Phys. 97: 013905.

42. Lagarkov AN, Osipov AV, Rozanov KN, Starostenko SN2005Microwave Composites Filled with Thin Ferromagnetic Films. Part I. Theory. Proc.

Symp. R: Electromagn.. Mater., 3rd Int. Conf. on Mater. Adv. Technol. (ICMAT 2005), Jul. 3−8, 2005, Singapore, 7477

43. O. Acher, S. Dubourg, 2008Generalization of Snoek's Law to Ferromagnetic Films and Composites. Phys. Rev. B 77: 104440.

44. Snoek JL1948Dispersion and Absorption in Magnetic Ferrites at Frequencies above 1 Mc/s. Physica 14207217

45. O. Acher, A. L. Adenot, 2000Bounds on the Dynamic Properties of Magnetic Materials. Phys. Rev. B 62: 11324.

46. Rozanov KN, Koledintseva MY, Drewniak JL2012A Mixing Rule for Predicting of Frequency Dependence of Material Parameters in Magnetic Composites J. Magn. Magn. Mater. 324: 1063−1066.

47. Lagarkov AN, Matytsin SM, Rozanov KN, Sarychev AK1998Dielectric Properties of Fiber-Filled Composites. J. Appl. Phys. 84: 3806−3814.

48. S. M. Matitsine, K. M. Hock, L. Liu, Y. B. Gan, A. N. Lagarkov, K. N. Rozanov, 2003Shift of Resonance Frequency of Long Conducting Fibers Embedded in a Composite. J. Appl. Phys. 94: 1146−1154.

49. Vinogradov AP, Machnovskii DP, Rozanov KN1999Effective Boundary Layer in Composite Materials. J. Communic. Technology Electr. 44: 317−322.

50. Lagarkov AN, Sarychev AK, Smychkovich YR, Vinogradov AP1992Effective Medium Theory for Microwave Dielectric Constant. J. Electromagn. Waves Appl. 611591176

51. Lagarkov AN, Sarychev AK1996Electromagnetic Properties of Composites Containing Elongated Conducting Inclusions. Phys. Rev. B 5363186336

52. Makhnovskiy DP, Panina LV, Mapps DJ, Sarychev AK2001Effect of Transition Layers on the Electromagnetic Properties of Composites Containing Conducting Fibres. Phys. Rev. B 64: 134205.

53. L. Liu, L. B. Kong, G. Q. Lin, S. Matitsine, C. R. Deng, 2008Microwave Permeability of Ferromagnetic Microwires Composites/Metamaterials and Potential Applications. IEEE Trans. Magn. 44: 3119−3122.

54. Han MG, Liang DF, Deng LJ2011Fabrication and Electromagnetic Wave Absorption Properties of Amorphous Fe79Si16B5 Microwires. Appl. Phys. Lett. 99: 082503.

Chapter 4

POLICRYPS COMPOSITE MATERIALS: FEATURES AND APPLICATIONS

R. Caputo[1], L. De Sio[1], A. Veltri[1], A. V. Sukhov[2], N. V. Tabiryan[3] and C. P. Umeton[1]

[1]LICRYL (Liquid Crystals Laboratory, IPCF-CNR), Center of Excellence (CEMIF. CAL) and Department of Physics, University of Calabria, Arcavacata di Rende, 87036 Cosenza, Italy

[2]Institute for Problems in Mechanics, Russian Academy of Science, Moscow 119526, Russia

[3]Beam Engineering for Advanced Measurements Company, Winter Park, Florida 32789, USA

INTRODUCTION

In recent decades, great attention has been devoted to the realization of electrically switchable holographic gratings in liquid crystalline composite materials. It has been shown, indeed, that devices based on holographic polymer dispersed liquid crystals (HPDLCs) are of low cost and can exhibit good diffraction efficiency (DE) [Margerum et al, 1992; Sutherland et al, 1996]. However, application oriented utilization of these devices is limited, in general, by their strong scattering of light, due to the circumstance that the droplet size of the nematic liquid crystal (NLC) component inside the polymer matrix is comparable to the wavelength of the impinging light. In this framework, we have recently proposed a new kind of holographic grating called POLICRYPS, made of polymer slices alternated to films of regularly aligned NLC. These structures do not present those optical inhomogeneities that are due to the presence of NLC droplets in usual HPDLC samples [Sutherland et al, 1994], and can therefore exhibit good optical characteristics, with values of the diffraction efficiency as high as 98%. This chapter is devoted to give an overview of the POLICRYPS as a composite material, along with a description of its main applications. After a short presentation of the structure, in terms of its fabrication process, we present the POLICRYPS as the device it was initially designed for: a switchable diffraction grating. We also demonstrate that, by suitably choosing the sample thickness and geometrical parameters, a POLICRYPS put perpendicular to the impinging light can behave as a switchable optical phase modulator, where the retardation between ordinary and extraordinary waves can undergo a fine electrical regulation. It is very interesting to show how by adding dye materials to the initial chemical mixture, necessary for obtaining POLICRYPS, can change the way we control its functionalities. In this case, such control can be obtained by using a laser beam of the right colour and power. This fact allows the realization of completely new applications like an optically controlled tunable beam splitter.

New intriguing sceneries open when using different materials to realize POLICRYPS. In particular, if a dye-doped cholesteric liquid crystal (instead of a nematic liquid crystal) is used, the POLICRYPS polymeric channels become mirrorless optical cavities where a distributed feedback (DFB) lasing effect (with a very low threshold) can be obtained. Another challenging opportunity is offered in case a tiny concentration of metallic nanoparticles is included in the initial POLICRYPS mixture. By doing so, we obtain a new device whose frequency spectrum is dependent on the probe light polarization. This last possibility is still in progress and is oriented to the realization of a POLICRYPS structure with meta-material properties.

THE POLICRYPS STRUCTURE

The morphology of the POLICRYPS is quite different from the HPDLC one. Optical microscope and scanning electronic microscope (SEM) investigations have shown that the structure consists of rigid slices of almost pure polymer alternated to films of almost pure NLC. The polymeric slices are well glued to two glasses that confine and contain the POLICRYPS. These slices represent a rigid frame that, somehow, 'stabilizes' the NLC component and, therefore, the whole sample. Separation interfaces between polymer slices and NLC films are quite regular and sharp; furthermore, there is convincing evidence that, at these interfaces, the NLC director is everywhere perpendicular to them, thus inducing a good, uniform alignment of the director in the whole NLC film standing between two polymeric slices. This circumstance represents one of the main features that determine the overall characteristics of the whole structure. The uniform and regular alignment of the director in the NLC films of the structure determines the main optical and electro-optical properties of the POLICRYPS. From the optical point of view, losses due to the scattering of the visible light (which is eventually brought to impinge onto the POLICRYPS) are reduced to less than 2%, thanks to the absence of droplets, which exist in HPDLC samples, with an average size comparable to the light wavelength and an arbitrary director alignment. From an electro-optical point of view, the fact that the NLC molecules are confined (and well aligned) in a uniform film, rather than in a small droplet, allows a suitably oriented electric field of the order of a few V/μm to uniformly 'reorient' the NLC director in a millisecond timescale. Afterwards, by suitably choosing the values of the refractive index of the polymer and the ordinary/extraordinary refractive index of the NLC, this director reorientation can be exploited to vary the spatial modulation of the refractive index of the POLICRYPS.

Fabrication Process and Set-up

The standard procedure that enables the realization of a good POLICRYPS structure exploits the high diffusivity of NLC molecules in the isotropic state, which avoids the formation and separation of the nematic phase (as NLC droplets) during the curing process [Caputo et al, 2000, 2004, 2007]. The main fabrication steps can be illustrated as follows. By means of a hot stage, a syrup of NLC, monomer and photo-initiator is heated up to a temperature which is above the nematic–isotropic transition point of the NLC component; the sample is then 'cured' with the interference pattern of a UV radiation. After the curing process has come to an end, the sample is brought below the isotropic–nematic transition point by means of a controlled, very slow, linear cooling down to room temperature. The experimental set-up exploits an active system for suppression of vibrations [De Sio et al, 2006, 2008a] and is presented in Fig. 1. An Ar-ion laser is the source of a single-mode radiation at the wavelength $\lambda_B = 351$ nm. The beam is broadened up to a diameter of about 25 mm by the beam expander BE, and divided into two beams of almost equal intensity by the beam splitter BS. These two beams overlap and give rise to the 'curing' interference pattern at the entrance plane of the sample cell S, whose temperature is controlled by the hot stage. Depending on the required nano/microscale dimensions of the structure, the spatial period of the interference pattern can be varied in the range $\Lambda = 0.2–15$ μm by adjusting the total interference angle $2\theta_{cur}$. A commercial, metal-coated, reflective diffraction grating (Edmund Optics) placed above the sample is used as a test element for the interferometric monitoring of vibrations. Part of each of the curing beams is reflected and diffracted by this grating. The set-up is adjusted to make the reflected part of one beam spatially coincident with the diffracted part of the second one. These two radiations are wave coupled by the test grating and their interference pattern is detected by an additional photodiode PD3. The signal of this photodiode is sent to a computerized active feedback system, which exploits a software that is based on a proportional–integral–derivative (PID) protocol; this drives a mirrorholder whose position can be controlled by a piezoelectric mechanism, used in feedback configuration. This control system has proved to be able to continuously compensate for changes in the optical path length due to vibrations as well as variations in environmental conditions such as room pressure, temperature or humidity; residual fluctuations are of the order of 6–7 nm, which correspond to the sensitivity of the piezo-system used.

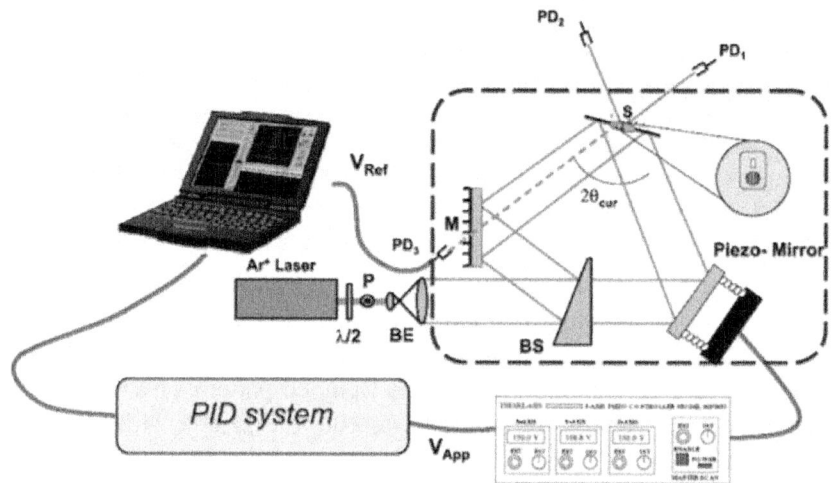

Figure 1: Optical holographic set-up for UV curing of gratings with stability check. P, polarizer; $\lambda/2$, half-wave plate; BE, beam expander; BS, beam splitter; $2\theta_{cur}$, total curing angle; M, mirrors; S, sample; PD1, first beam photo-detector; PD2, second beam photo-detector; PD3, diffracted/reflected beam photo-detector. In the insertion the reference grating is shown (put immediately below the sample area) which enables the stability check.

THE POLICRYPS GRATING/PHASE MODULATOR

POLICRYPS as a High Quality, Switchable, Diffraction Grating

The basic device that can be realized by using electrically switchable holographic gratings in liquid crystalline composite materials is an electro-optical switch [Sutherland et al, 1994]. Such a device should, in principle, completely diffract or transmit an impinging light beam, depending on the application of an external voltage (Fig. 2).

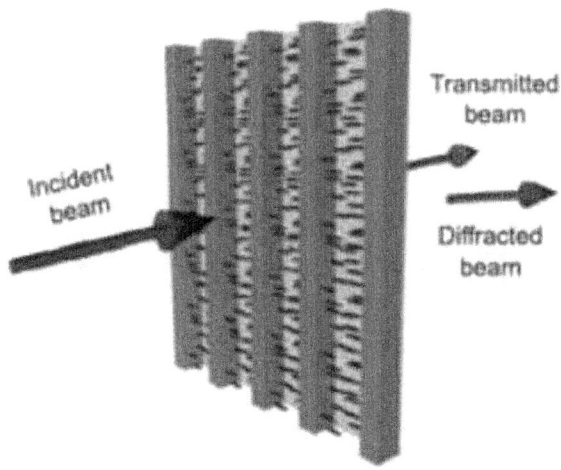

Figure 2: Sketch of a POLICRYPS grating in transmission configuration.

PDLCs have been actively utilized in the past in order to realize working prototypes of this kind of device; unfortunately, they still show issues that affect their performances. One of the main reasons that brought us to design POLICRYPS systems was the possibility of overcoming most of these issues. In the following, we report the results of an experimental comparison between an HPDLC and a POLICRYPS grating, in order to put into evidence how microscopic features of the structure can influence the overall performance of the macroscopic device. We have realized a standard HPDLC and a POLICRYPS grating, both with a fringe spacing Λ=1.5 μm. Sample cells, 16 μm thick, made with indium tin oxide (ITO)-coated glass slabs, were filled with the same initial chemical syrup. This was prepared by diluting the NLC 5CB (Merck, ≈30 wt%) in the prepolymer system Norland Optical Adhesive NOA-61. The POLICRYPS grating was cured by a total UV intensity of 11 mW/cm², acting on the sample for τ≈1000 s at high temperature (e.g. above the nematic–isotropic transition point of the 5CB liquid crystal), these being the optimal conditions for achieving a high diffraction efficiency and a morphology of good quality [Caputo et al, 2001]. Almost the same UV intensity and curing time proved also to be adequate for the realization of the PDLC grating, but in this case the sample was cured at room temperature. In order to explore the performances of both gratings, we used a weak (P≈1mW) He–Ne laser radiation (λ_R=633 nm), with its angle of incidence adjusted for satisfying the Bragg condition for the first-order diffracted beam. With the aim of performing a comparison in the same experimental conditions, before starting the curing process of each sample, we measured the intensity Iin of the impinging probe beam (before the sample) and the transmitted intensity Itr. Then, once the curing process had been completed and the UV light switched off, we measured both the intensity I0 of the zero-order (direct transmitted) probe beam and the intensity I1 of the first-order diffracted probe beam. In this way we were able to calculate the zero-order transmittivity $T_0 = I_0/I_{in}$, the first-order transmittivity $T_1 = I_1/I_{in}$, the total transmittivity $T_{tot} = T_0 + T_1$ and the first-order diffraction efficiency, which is usually calculated as $\eta_1 = I_1/I_{tr}$. During all the experiments, the intensity of the probe beam was maintained at a fixed value (the value of the initial impinging intensity before the curing process started). We measured the first-order diffraction efficiency at room temperature both for POLICRYPS and HPDLC gratings, obtaining $\eta_{POLICRYPS}^1 = 88\%$ and $\eta_{HPDLC}^1 = 41.2\%$. . We stress that the value of η POLICRYPS 1 is not the highest that we can get since, by using other POLICRYPS gratings (not involved in comparisons with HPDLC ones), we have obtained $\eta_{POLICRYPS}^1$ values as high as 98%. The electro-optic response of the two gratings was investigated by exploiting a low frequency (500 Hz) square-wave voltage, and results are reported in Fig.3.

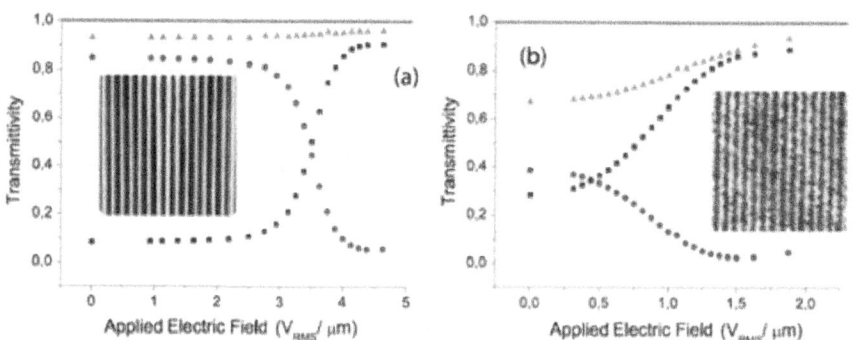

Figure 3: Dependence on applied voltage of the zero-order transmittivity T_0 (squares), first-order transmittivity T1 (circles) and total transmittivity T_{tot} (triangles) for (a) a POLICRYPS grating and (b) an HPDLC grating at room temperature. Error bars are of the order of the dot size. The pictures in the inset show respectively a typical POLICR-YPS and HPDLC grating morphology observed under a polarizing optical microscope

Figure. 3a represents the switching curve of the POLICRYPS grating: the behaviour of the firstorder transmittivity T_1 (circles), zero-order transmittivity T_0 (squares) and total transmittivity T_{tot} (triangles) is reported versus the root mean square applied electric field.

Table 1: Measured values of the switching times for a POLICRYPS and an HPDLC grating obtained from the same initial mixture

	$\tau_{\text{fall}}\ (\text{ms})$	$\tau_{\text{rise}}\ (\text{ms})$
POLICRYPS	1.12 ± 0.03	0.88 ± 0.03
HPDLC	10.53 ± 0.18	1.36 ± 0.04

It is worth noting that T is only slightly lower than 1 and remains approximately the same for all values of the applied field; this indicates that the grating exhibits negligible scattering losses. The situation is quite different for the HPDLC grating (Fig. 3b): the total transmittivity is well below 1 and increases as the applied field increases. We also note that the switching efficiency $h_{on} \cong \frac{T^1_{on} - T^1_{off}}{T^1_{on}}$, where T^1_{on} and T^1_{off} are the first-order transmittivities in the switch-on and switch-off condition respectively, is almost the same (93.3%) for both gratings. Where the switching fields are concerned, the first diffracted beam is almost completely switched off by a field of about $1.5\,\text{V}/\mu\text{m}$ applied to the HPDLC grating, while a value of about $4.3\,\text{V}/\mu\text{m}$ is needed to obtain the same effect in the POLICRYPS one. This particular difference can be due to the average size of NLC droplets in the HPDLC; evidently, this size is large enough to enable low switching fields. This is confirmed by the values of the switching times shown in table 1: both the rise and fall times of the HPDLC grating are longer than those of the POLICRYPS; this suggests a very large average size of PDLC droplets. Here, we stress that the electro-optic behaviour shown in Fig. 3a, and its noticeable difference with the one of Fig. 3b, represents the best evidence of the good performances of POLICRYPS gratings; indeed, people working with HPDLCs of nanosized droplets also find for these materials behaviours that are comparable to the one shown in Fig. 3a, but for values of the switching fields which are about four-fold higher [Lucchetta et al, 2003].

Policryps as an Optical Phase Modulator

The preferential orientation and the good alignment assumed by the molecular director n of the LC material within a POLICRYPS structure recently suggested a possible use of these systems as switchable phase modulators [De Sio et al, 2008b]. Examples of such devices are already present in literature. A basic embodiment is obtained by enclosing a NLC with a positive dielectric anisotropy in a cell made of two ITO-coated glasses, treated to give a planar alignment to the NLC director. Since the liquid crystal is birefringent, light with wavelength λ, propagating through the structure, is separated into an ordinary and an extraordinary component. If L is the thickness of the sample and Δn_{LC} indicates its birefringence, the phase difference δ_{LC} between these two waves, measured at the exit of the sample, depends on the value of

Δn_{LC}: $_{\delta LC}=2\pi L\Delta n_{LC}/\lambda$. By applying an external electric field E with direction perpendicular to the glass slabs of the cell, n tends to reorient along the same direction as E, thus producing a change in the phase difference. However, this simple device presents some drawbacks. The orientation of n is, indeed, sensitive to temperature changes [de Gennes, 1993]. This can represent a serious limit for an eventual device when the power of the impinging radiation is high. Moreover, the switching times of such devices are usually quite long (2–8 ms), thus limiting the field of possible applications. In order to overcome the above-mentioned problems, the NLC layer is often stabilized by means of polymeric chains [Wu et al, 2004]; their presence improves the response times of the device but, unfortunately, drastically increases the operating voltages (due, probably, to the torque exerted by the polymer on the nematic director). Moreover, due to the irregularity of morphology induced by the presence of polymeric chains, visible light is strongly scattered. Therefore, these systems are suitable only for wavelengths in the infrared range. Several features of POLICRYPS structures make them an attractive alternative to the discussed system. First of all, they exhibit limited scattering losses when illuminated by visible light. Second, the polymer slices confine and stabilize the NLC molecules, thus also influencing their alignment, and third, POLICRYPS structures can be driven by low voltages exhibiting short switching times. We expect that the better the alignment of the NLC director in the nematic layer of the POLICRYPS, the higher the value of Δn_{LC}; then, the phase retardation introduced by the grating will depend on the angle that the polarization vector of the impinging light forms with the nematic director within the LC layers Fig. 4.

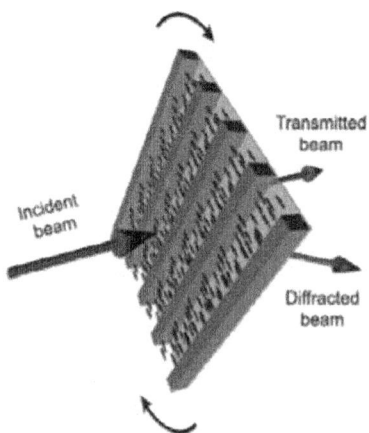

Figure 4: Sketch of a POLICRYPS grating in transmission configuration used as a phase modulator.

It is important to underline that, because of the diffractive nature of a POLICRYPS structure, the light impinging on the device will not only experience a phase modulation but will also undergo a dichroic absorption as explained in [Caputo et al, 2010]. This double behaviour can be taken into account by considering the POLICRYPS, in terms of the Jones matrix formalism, as both a retardation plate and a dichroic absorber. This is done by multiplying the Jones Matrix of the generic phase retarder by a new matrix L given by:

$$L = \begin{pmatrix} H & 0 \\ 0 & V \end{pmatrix}$$

(1)

where H and V parameter values depend on the considered material and can reflect a broad range of situations. The POLICRYPS used for experiments as a phase modulator has a thickness L=3.03μm and a fringe spacing Λ=1.22μm. In order to check the phase retardation properties of this structure, we used the experimental set-up reported in Fig. 5. The POLICRYPS is put between a polarizer P and an analyzer A, with its optical axis oriented at an angle θ=π/4 with respect to the first polarizer; in this position, the field components have the same amplitude $(E_\parallel = E_\perp)$ and the sample introduces the maximum retardation. During experiments, the position of the sample remains fixed while the analyzer is rotated (in steps of 10°) around the axis of propagation of the probe light (z axis in Fig. 5). We define β as the angle between directions of analyzer and incident polarization (therefore β=0 when the analyzer A is parallel to the polarizer P). If we indicate with I_{inc} the intensity of the impinging beam, by means of eqs (2) and (3) (derived in [Caputo et al, 2010]), it is possible to calculate the complex electric field $\tilde{E}_{out}(\beta)$ and hence the intensity $I_{out}(\beta)$ of light transmitted by the analyzer A in our experimental geometry.

$$\tilde{E}_{out}(\beta) = \frac{\sqrt{2}}{2}\sqrt{I_{inc}} \begin{pmatrix} -He^{i\frac{\delta}{2}}\sin^2\beta - Ve^{-i\frac{\delta}{2}}\sin\beta\cos\beta \\ He^{i\frac{\delta}{2}}\sin\beta\cos\beta + Ve^{i\frac{\delta}{2}}\cos^2\beta \end{pmatrix}$$

(2)

$$I_{out}(\beta) = \tilde{E}_{out}(\beta) \cdot \tilde{E}_{out}^*(\beta) = \frac{I_{inc}}{2}\left[H^2\sin^2\beta + V^2\cos^2\beta + HV\sin 2\beta\cos\delta \right]$$

(3)

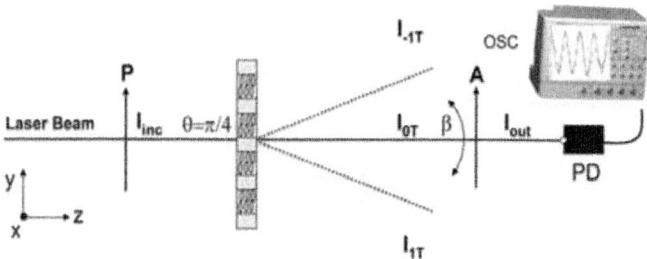

Figure 5: Experimental geometry utilized for measuring the intensity transmitted by the system composed of a birefringent/dichroic sample put between two polarizers. P polarizer, A analyzer, I_{inc} total incident intensity, I_{out} output intensity, I+T and I±1T zeroth and first order transmitted intensities, respectively. θ is the angle between the light polarization direction (y axis) and the grating optical axis (laying in the xy plane), PD Photo-detector, OSC oscilloscope. The probe beam is from a He-Ne laser at the wavelength λ=632.8 nm. S is the POLICRYPS sample

$$H = \sqrt{\frac{2I_{out}\left(\beta = \pi/2\right)}{I_{inc}}}$$

(4)

$$V = \sqrt{\frac{2I_{out}\left(\beta = 0\right)}{I_{inc}}}$$

(5)

While the phase retardation δ introduced by the sample can be calculated as:

$$\cos\delta = \frac{1}{HV}\left[\frac{2I_{out}\left(\beta = \pi/4\right)}{I_{inc}} - \frac{H^2 + V^2}{2}\right]$$

(6)

By substituting obtained data in eqs. (4), (5) and (6), we obtain: H=0.727, V=0.406 and δ=1.26rad. In Fig. 6, the experimental value of Iout as a function of the angle β (crosses) is compared with the theoretical behavior predicted by eq. 3 (solid line). The different values of H and V show that, even at normal incidence, the diffraction efficiency of the POLICRYPS grating is significant. As for the birefringence of the structure, the obtained value of δ yields Δn=0.042. By considering that the periodicity of the grating is much larger than the probe wavelength we can exclude that this considerably high value is due to form birefringence and hence to the geometrical features of the grating. We are confident, instead, that this value indicates that the stabilizing and confining action exerted by polymer slices on the NLC molecules has a direct influence on their alignment.

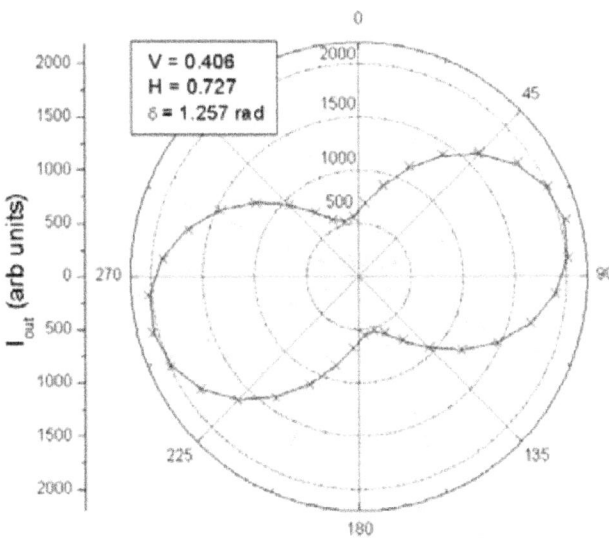

Figure 6: Behaviour of the intensity transmitted by the analyzer put after a POLICR-YPS grating as a function of the angle β between the electric field of the impinging wave and the axis of the analyzer itself. Two segments in the graph evidence output intensity values for the analyzer positions β=0 and β=π/2 respectively

Tunability of The Phase Retardation

In order to verify the functionalities of POLICRYPS as a tunable phase retarder, we prepared another sample whose thickness had a wedge shaped profile [Caputo et al, 2011a]. As discussed above, by applying an external electric field we can change the birefringence and hence the phase retardation introduced by the structure. However, the phase retardation also depends on the thickness of the layer in which the light propagates. This explains the choice of a wedge shaped cell: by combining the application of the electric field and the possibility to shift the sample to get the desired thickness, it is possible to achieve a very fine tunability for our device. The realized wedge-shaped structure has a thickness varying in the interval (3.00÷5.00 µm) and has been experimentally characterized by means of the setup shown in Fig. 5. The check of the electro-optical tunability of the sample birefringence, has been performed by probing the fabricated sample in the area corresponding to a thickness L=4.35µm. This and other thickness values have been measured before filling the cell by means of an Agilent spectrophotometer and considering the cell as a Fabry-Perot etalon. The applied electric field is a bipolar square wave with frequency ν=1 kHz and a peak-to-peak amplitude varying in the interval (0÷9 Volts/µm). Measurements of the intensity $I_{out}(β)$ of light transmitted by the analyzer A have been performed by changing β in the interval 0≤β≤2π, for different values of the applied electric field. Obtained results show that the application of an electric field produces a tuning action of the phase retardation from 1.64rad to 1.07 rad. In each curve of Fig. 7, dots represent experimental values whereas solid lines indicate theoretical predictions; it is evident that the agreement is very good. The plot of both the birefringence value Δn (red dots) and the phase retardation δ (blue dots) of the structure, calculated considering a thickness L=4.35μm, are reported in Fig. 8 as a function of the applied electric field. Phase retardation variations yield, in this case (λ=632.8nm), to a birefringence value varying in the interval (0.024÷0.038). The phase retardation/birefringence properties of our POLICRYPS structure can be also varied by shifting the probed area of the sample along the wedge direction. Several positions have been probed. Experimental results and corresponding theoretical curves are shown in Fig. 9.

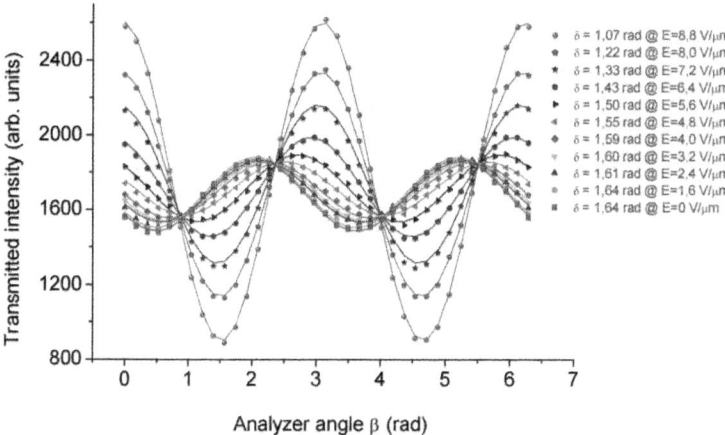

Figure 7: Behavior of the intensity transmitted by the analyzer put after the POLICR-YPS grating obtained by changing the amplitude of the applied electric field. For each amplitude, the output intensity has been measured by varying β between 0 and 2π. Solid lines are theoretical fits while dots represent experimental data. Experimental errors are of the order of the dot size

Also in this case, results confirm the possibility of tuning the phase retardation at will by just probing the sample in the position where it has the right thickness.The thickness value L=4.10μm corresponds to a phase retardation δ=1.55rad (orange curve in Fig. 9) which is close to the condition of quarter wave plate for the He-Ne laser wavelength. This curve is almost constant for every β angle, as expected for this particular value of the phase retardation.

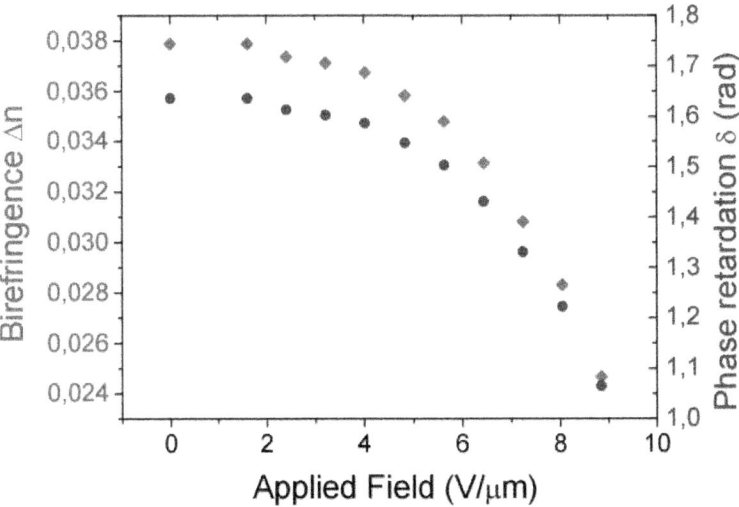

Figure 8: Plot of birefringence Δn (red dots) and phase retardation δ (blue dots) of the POLICRYPS structure versus the amplitude of the applied electric field.

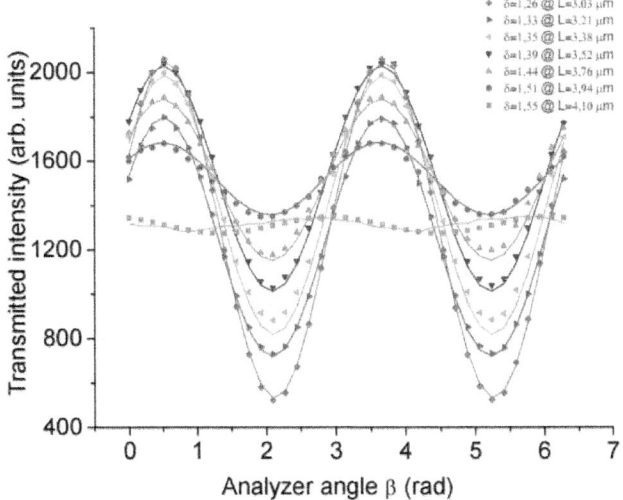

Figure 9: Behavior of the intensity transmitted by the analyzer put after a POLICR-YPS grating obtained by shifting the sample along the wedge direction and probing it in areas with different thickness. For each thickness, the output intensity has been measured by varying β between 0 and 2π. Solid lines are theoretical fits while dots represent experimental data. Experimental errors are of the order of the dot size.

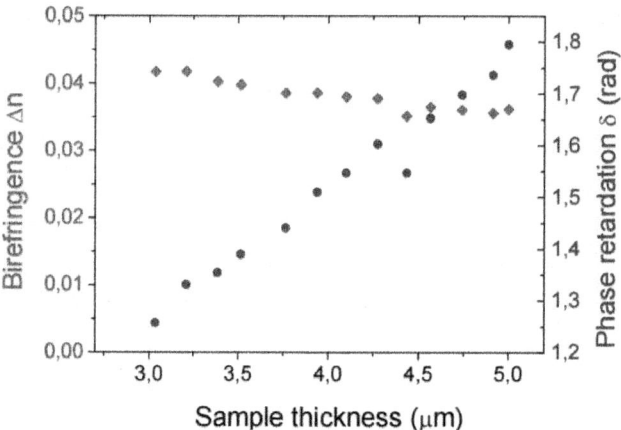

Figure 10: Plot of birefringence Δn (red dots) and phase retardation δ (blue dots) of the POLICRYPS structure measured by shifting the sample along the wedge direction and probing it in areas of different cell thickness.

This result confirms the possibility of finely tuning the phase retardation introduced by the structure by playing both with the amplitude of the applied electric field and the position of the sample for finding the area corresponding to the optimal thickness. Birefringence and corresponding phase retardation values, measured by shifting the sample along the wedge direction, are reported in Fig. 10. We can notice that, by increasing the thickness L in the interval (3.0÷5.0μm), Δn remains almost constant, as expected if we consider that the POLICRYPS grating exhibits a quite homogeneous morphology. On the contrary, the phase retardation δ shows a linear increase with values varying in the interval (1.25÷1.80 rad).

PHOTORESPONSIVE POLICRYPS STRUCTURES

As discussed above, a fundamental advantage provided by holographic structures containing liquid crystal materials is the possibility of tuning their optical properties by applying external electric or thermal fields. Some years ago, Tondiglia et al. have proposed another fascinating possibility: the use of light for switching the optical properties of gratings [Bunning et al, 2000]. Indeed, azobenzene liquid crystals enable to access, optically and isothermally, a nematic to isotropic (NI) transition that changes the refractive index of the liquid crystal films, thus modifying the refractive index modulation of the whole structure. The exploited mechanism is that, upon UV (λ=360 nm) irradiation, azo-LC molecules undergo a conformational change (from rodlike trans to cis) which drives the LC through an isothermal NI phase transition; this process can be driven in the reverse direction by converting the cis-azobenzene moieties back to their rodlike trans state via exposure to a radiation of a suitable wavelength [Tsutsumi & Ikeda, 1995]. The decision to adopt these materials in the initial POLICRYPS mixture brought to the realization of the socalled azo-POLICRYPS: optically controlled POLICRYPS structures. Samples have been realized by means of the typical setup for POLICRYPS fabrication (reported in Fig. 11). Preliminary attempts, performed on different sample cells, have shown that the best performances are exhibited by the one of L=11.4μm thickness, with a grating pitch Λ=1.6μm, this value being such that the highest diffraction efficiency is obtained with the actual cell thickness. Samples have been prepared by using the following mixture: 25 wt % of NLC (E7 by Merck), 5 wt % of azo-LC (1005 by BEAM Co.), and 70 wt % of monomer (NOA61 by Norland). A qualitative characterization, made with an optical microscope, shows that this azo-POLICRYPS exhibits a stable structure, made of alternated layers of pure polymer and pure LC, with the apparent absence of PDLC droplets (inset of Fig. 11).

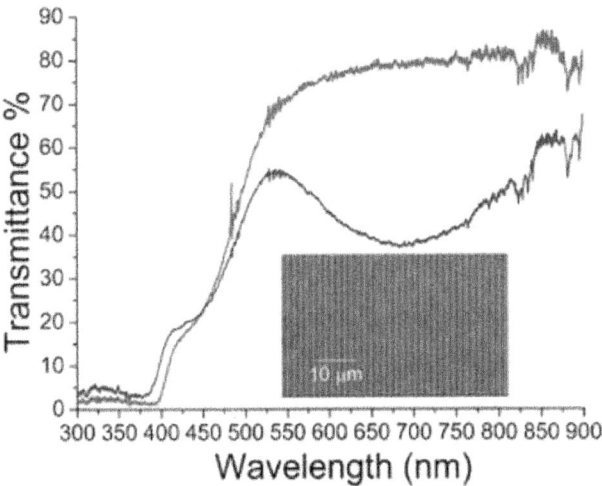

Figure 11: Transmission spectra of both the uncured mixture (red curve) and the realized grating (blue curve). Inset: photo of a POLICRYPS grating taken with a 20x objective equipped Olympus microscope

Transmission spectra reported in Fig. 11 have been obtained with the aid of a fiber optic spectrometer, using light beams of arbitrary polarization at normal incidence. The mixture exhibits a high absorption in the λ=300–400 nm range, while the grating gives rise to a relative minimum in the transmission around λ=650 nm.

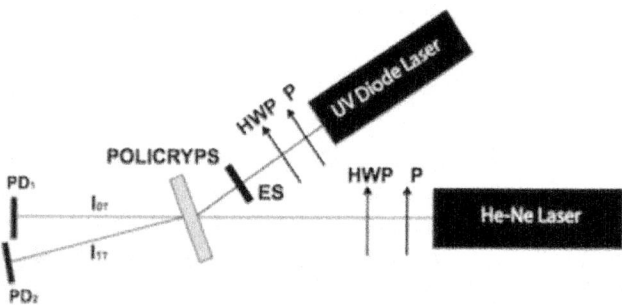

Figure 12: Experimental setup for the observation of all-optical processes in azo-POLICRYPS. PD1,2: photodetectors; HWP: half-wave plate; P: polarizer; ES: electronic shutter

Thus, the setup utilized for the characterization of the sample (Fig. 12) makes use of an UV diode pump laser emitting at $\lambda=409$ nm (in the high absorption range of the mixture spectrum) and a He–Ne probe beam of $\lambda=633$ nm (in the range that is of high efficiency for the grating); this beam impinges at the Bragg angle $\theta=11.5°$. For experimental simplicity, we have used an unfocused pump beam with a power of only 4.4 mW, which exhibits an oval shape on the sample of about 2-3 mm2. Fig. 13 shows that a pump irradiation of duration $\tau=20$ s, operated by opening the electronic shutter (ES), reduces the diffraction efficiency of the azo-POLICRYPS grating to less than 75% of its initial value in a time of a few seconds; a slow increase toward the initial value is then observed when the shutter closes the pump beam.

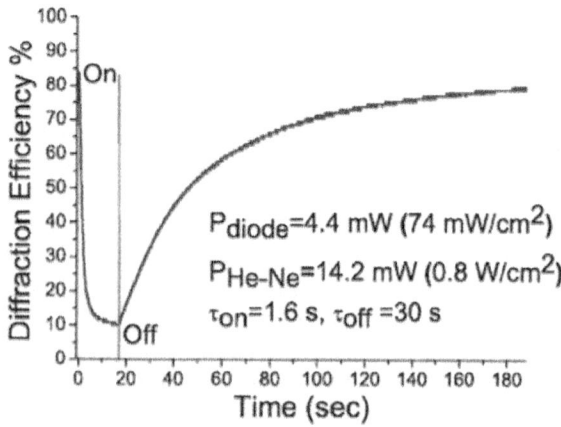

Figure 13: Dynamics of the diffraction efficiency of the probe first order diffracted beam. The diffraction efficiency of the grating is de- fined as the diffracted intensity divided by the sum of the diffracted and transmitted intensities.

We have investigated this dynamics by varying the pump power from 0.1 mW to 14.6 mW, while the probe power is kept constant: (inset of Fig. 14).

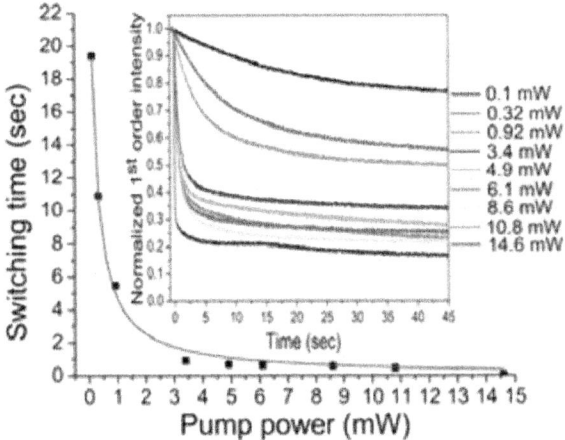

Figure 14: Response time of the azo-POLICRYPS grating vs the power of the pump beam; the error bar is of the order of the square dimension. Diffraction efficiency values in the inset are normalized to the initial ones.

The switching time (sw) of each curve is reported in Fig. 14 as a function of the impinging pump power P: when this power exceeds a minimum value Pmin≅0.1 mW (below which the effect is very small and it is almost impossible to get a good signal), data are well fitted by the negative exponential red curve,

$$\tau_{sw} = \tau_0 e^{-(P-P_{min})/P_0}$$

(7)

where τ_0=19.4 s and P_0=1.8 mW; this behavior can be explained by assuming the rate of concentration of photoisomerized azo-LC molecules proportional to the impinging intensity. Taking into account that in the actual experiment the pump beam is not focused, we foresee that much shorter switching-off times can be achieved in systems irradiated by a focused beam, where a high power density is obtained with power levels even lower than the actual ones.

azo-POLICRYPS as an Optically Controlled Beam-Splitter

The possibility to realize an optically controlled POLICRYPS structure is of particular interest for applications, since fast light responsive devices represent an innovative way to realize an on-chip technology. A fundamental element of an optical set-up is the beamsplitter. This element splits an incident light beam into two or more beams, which may or may not have the same intensity. These devices are usually passive in the sense that the intensity of the splitted beams is fixed or it can be eventually varied by changing the angle of incidence of the light impinging on the device. POLICRYPS diffraction grating can combine the capability of "dividing" beams (typical of periodic structures) with the effects of an adjustable birefringence, typical of nematic liquid crystals, which influences the intensities of the diffracted beams.

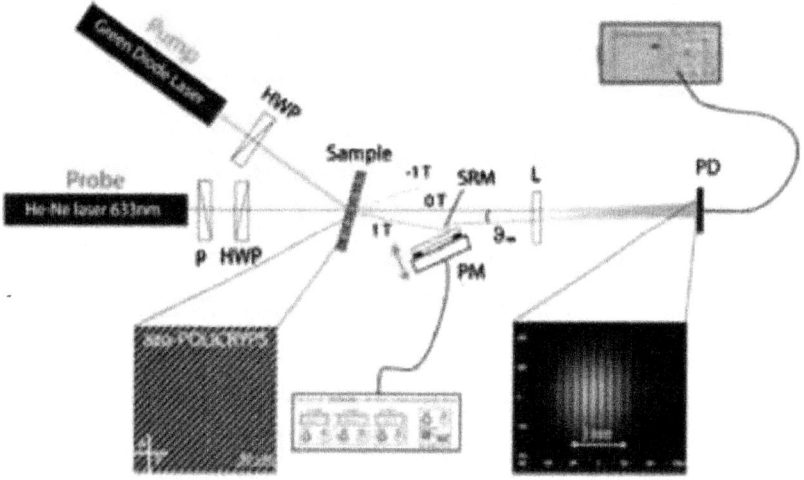

Figure 15: All-optical OBS and interferometer setup: P, polarizer; HWP, half-wave plate; SRM, semireflective mirror; θint, interference angle; PM, piezomirror; PD, photodetector; L, lens.

By using an azo-POLICRYPS as an optical beam splitter (OBS) it is possible to obtain a tunable optical element, where we can choose, at will (turning a knob), the portion of initial intensity which remains to the transmitted beam and the amount which is transferred to the splitted (diffracted) one. For the experiment, we have fabricated an azo-POLICRYPS structure whose geometrical parameters are L=6.95μm in thickness and Λ=1.57 μm in fringe spacing; according to Kogelnik's theory, [Kogelnik, 1969] this grating operates in the Bragg regime with a characteristic parameter $\rho = \Lambda_2/\lambda L = 0.56$ at $\lambda = 0.633$ nm. The experimental setup utilized to exploit the azo-POLICRYPS as a finely adjustable, optically controlled OBS is reported in Fig. 15. The impinging probe light is split into two beams (the transmitted and the diffracted orders, 0T and 1T respectively) by the azo-POLICRYPS grating. These are recombined in Mach–Zehnder interferometer geometry; this part of the setup is actually used to monitor the functionality of the OBS. The diffraction efficiency change of the azoPOLICRYPS is driven by an external pump source green diode laser. On application of the pump laser, the index contrast of the grating vanishes and the structure becomes transparent to the impinging probe light. Fig. 16a shows the evident change in the diffraction efficiency induced by switching ON the pump green light.

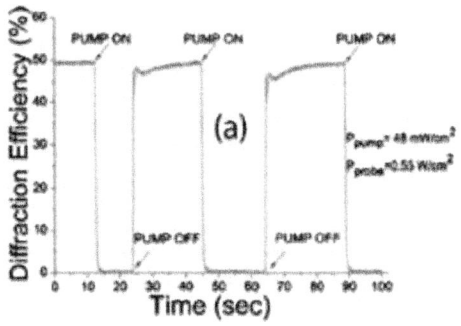

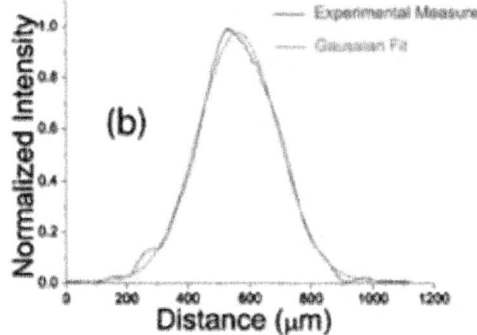

Figure 16: (a) Reversible and repeatable changes of the azo-POLICRYPS diffraction efficiency induced by a pump green light. Power density values are reported in the figure; (b) Typical intensity profile of the transmitted intensity trough the structure

The switching response of the azo-POLICRYPS OBS is detected by using a sequence of ONOFF pump beam irradiance P_{pump}=48 mW/cm2 while the intensity of the probe red beam is kept ON at all times P_{probe}=0.55 W/cm^2.

As for the quality of the transmitted (0_T) and diffracted (1_T) beams outgoing from our OBS, we have detected their transverse intensity profiles with a charge-coupled device CCD camera for different increasing values of the incident probe power; we had the evidence that the grating does not modify the typical Gaussian shape of both beams, which remains almost unchanged for any value of the impinging probe power density in the range 0.1 to 0.7 W/cm^2; Fig. 16b shows the Gaussian profile of the transmitted beam detected for the value Pprobe=0.7 W/cm^2. The ratio $R=I1_T/I0_T$ of the intensities of 1_T and 0_T beams is related to the diffraction efficiency of the azo-POLICRYPS through the equation

$$\eta = \frac{I_{1T}}{I_{0T} + I_{1T}} = \frac{R}{1+R} \qquad (8)$$

For the aim of the actual work, the polarization of the probe beam and its incident angle have been adjusted to obtain a maximum diffraction efficiency value max=50%, that is to say R_{max}=1, when the pump beam is off [De Sio et al, 2010]. In order to characterize and exploit the azo-POLICRYPS as a variable OBS, we have used the interferometer setup reported in Fig. 15. The interference pattern reported in the dark insets of Fig. 15 produced by overlapping 0T and 1T beams is monitored by means of the detector PD, which is provided of a small aperture 500 μm on top of the active area. The pattern periodicity can be easily controlled by varying the orientation of the semireflecting mirror, thus the angle θint. In our experiment, θint was relatively small 0.04°, and the scale is reported in the same dark inset of Fig. 15. The fringe visibility, defined as v=$(I_{max}-I_{min})/(I_{max}+I_{min})$, (where Imax and I_{min} are the measured maximum and minimum intensity values of the interference pattern) strongly depends on R. Indeed, for our geometry, it is easy to see that

$$v = \frac{2\left(I_{0T}I_{1T}\right)^{1/2}}{I_{0T} + I_{1T}}|\gamma| = \frac{2R^{1/2}}{1+R}|\gamma|$$

$$(9)$$

Here, γ (the degree of coherence of the two beams [Yariv, 1989] is related to the difference Δl of the optical path lengths of the two beams and to the coherence length lc of the probe laser beam, which in our case is of the order of 10 cm (HRP050 Thorlabs). We can assume that Δl does not exceed few micrometer even when the piezomirror (PM) is shifted back and forward of few micrometer, therefore $|\gamma|=(1-\Delta l/l_c)\approx 1$. Relating η to R by the equation R=η/(1−η) and substituting it into eq. (9) we obtain:

$$v = 2\left(\frac{\eta}{1-\eta}\right)^{1/2}$$

$$(10)$$

Since in our azo-POLICRYPS grating η varies with the impinging pump power P_{pump}, we have investigated the behavior of our tuneable OBS by detecting the fringe visibility v versus P_{pump}. Measurements have been performed by applying a linear voltage to the piezomirror PM included in the interferometric part of the setup of Fig. 15. In this way, we were able to modify the optical path length of one of the two arms, thus allowing a scrolling of the fringe pattern on the PD and a measurement of I_{max} and I_{min} values, without shifting the PD from the top of the impinging Gaussian beams. Indeed, a linear movement of the piezomirror in the direction normal to the mirror plane corresponds to a shift of the fringe pattern along a direction parallel to the PD surface; therefore, the output signal from the PD exhibits the sinusoidal behavior shown in Fig. 17.

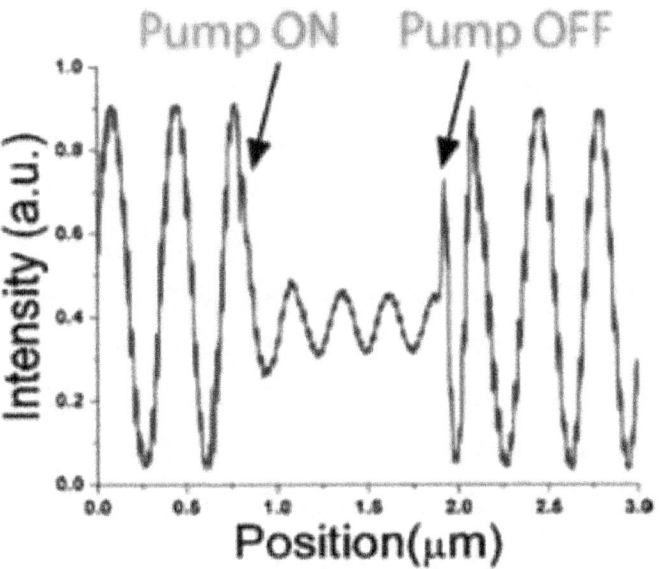

Figure 17: Intensity profile of the interference pattern vs the piezomirror position. A reversible change of oscillation amplitude, obtained by switching ON and OFF the external pump can be well observed.

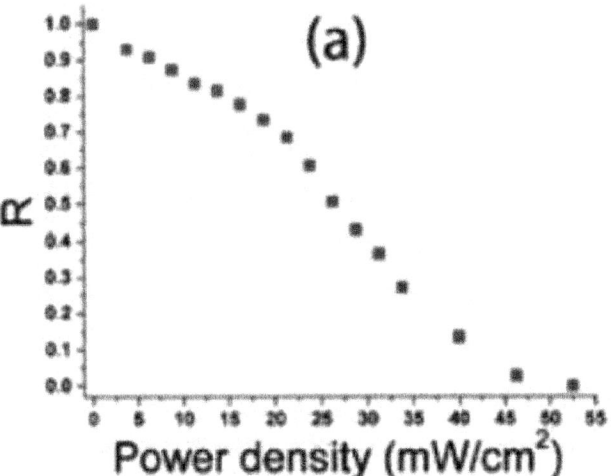

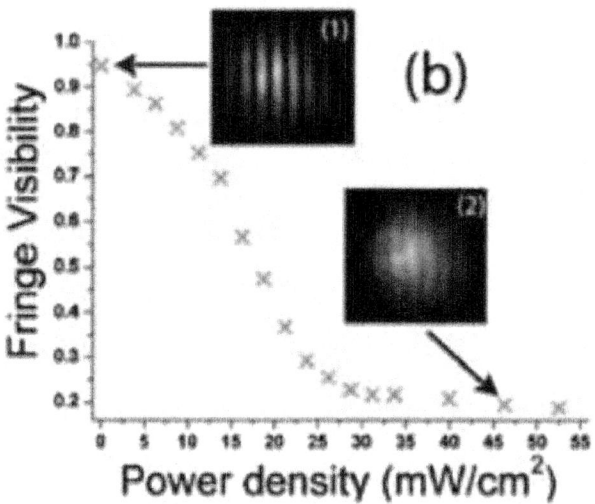

Figure 18: Beam splitting (a) and fringe visibility (b) vs the pump power density. Interference pattern acquired with a CCD camera is reported for v=0.94 (1) and v=0.2 (2). Experimental errors are the order of dot and cross dimensions

The amplitude of the sinusoidal modulation is strongly attenuated when irradiating with a green pump laser (P_{pump}=48 mW/cm2) over the spot of the red light; oscillation amplitude is restored to its initial value in some milliseconds by turning off the external pump. The behavior of v versus fine variations in P_{pump} is reported in Fig. 18b, along with measured values of R. Curves can be explained by considering that the rate of the trans–cis isomerization process depends on the number of excited molecules; therefore, the rate of concentration of photoisomerized azo-LC molecules is proportional to the pump power density. This phenomenon directly affects R and therefore v, which varies from 0.94 to 0.20. Fig. 18a shows that R values can be finely adjusted between 1 (transmitted and diffracted beams of the same intensity) and 0 (no diffracted beam, the whole impinging intensity is transmitted).

As for the measured v values, following eq. (9), also v should vary between 1 (when R=1) and 0 (when R=0). The observed discrepancy (0.92 instead of 1 and 0.2 instead of 0) can be explained by taking into account that, due to the birefringence of the grating [Caputo et al, 2010] the transmitted beam is elliptically polarized, with an ellipticity of the order of a/b≈10/1 where a and b are the major and minor axes of the polarization ellipse, respectively. The weak component polarized perpendicularly to the diffracted field is responsible for the small discrepancy between measured and predicted v values.

THE POLICRYPS AS AN ARRAY OF OPTICAL RESONA-TORS

A new intriguing scenario of applications emerges if we explore the possibility of obtaining a lasing action in POLICRYPS structures. Indeed, in recent years, many efforts have been spent in research for the realization of lasing devices based on organic systems: good candidate materials for achieving this result are cholesteric liquid crystals (CLCs). It is well known that CLC materials possess a helical periodic superstructure which provides a 1D spatial modulation of the refractive index [de Gennes, 1993]. This system behaves as a photonic band gap (PBG), i.e. it exhibits a window in the electromagnetic spectrum where wave propagation is forbidden. This is due to a mechanism known in literature as distributed feedback (DFB), and has the consequence that the system behaves as a mirrorless optical resonator. If the CLC material is doped with fluorescent guest molecules, a gain enhancement of the radiation, propagating in the structure, is possible. Kogelnik and Shank [Kogelnik & Shank, 1971] were the first to report laser action in mirrorless periodic Bragg DFB structures, while laser action in chiral liquid crystals was predicted by Goldberg and Schnur [Goldberg & Schnur, 1973]. There are many advantages in using POLICRYPS as a host structure for dye doped CLC helices. The sharp and parallel channels of POLICRYPS can behave as an array of optical resonators, each of them working as a microlaser. The length of the single channel is not limited by the sample geometry; in principle the single cavity can be several centimetres long, thus containing thousands of periods of the CLC helices. At the same time, its volume can be reduced at will by changing the periodicity of the structure. Optical resonators with these two features present a high quality factor Q and correspond to very efficient microcavity lasers. Such an array of microlasers has been experimentally realized in a POLICRYPS structure [Strangi et al, 2005]. A slightly different chemical mixture was used: a small amount (0.7 wt%) of Irgacure 2100 and Darocur 1173 photoinitiators (1:1 wt%, Ciba Specialty Chemicals) was used to reinforce the polymeric network and a 0.09 wt% of pyrromethene dye (Exciton) was added, which represented the gain medium of our system. Other components were 29.9 wt% BL088 cholesteric liquid crystal (Merck), and 69.3 wt% of NOA-61 monomer (Norland). The mixture was introduced by capillarity between ITO-coated glass plates separated by 13.5 μm thick mylar spacers. The sample was then prepared by following the typical recipe for obtaining POLICRYPS. The only difference is that the curing temperature was sensitively higher in order to bring the CLC material in the isotropic phase during curing. At the end of the whole process, an almost complete phase separation was obtained, giving rise to helixed liquid crystal channels periodically separated by polymer walls. A scanning electron microscopy analysis of the sample showed a periodicity of 5 μm with the microcavity width of about 1.5 μm. The system was optically pumped with the second harmonic ($\lambda = 532$ nm) of a Nd:YAG laser. The laser beam was focused onto the sample by means of a cylindrical lens ($f = 100$ mm) and linearly polarized perpendicularly to the microchannels. The long axis of the section was oriented perpendicularly to the orientation of the polymeric walls; therefore, the profile obtained (long axis of approximately 5 mm) ensured the simultaneous excitation of multiple microchannels. Above a certain pump power, stimulated emission was achieved, emerging from the microcavities in a direction parallel to the glass plates and along the microchannels. At their end highly sensitive emission measurements were performed in a restricted cone angle of about 0.1 rad. The sketch in Fig. 19 shows this lasing scenario of the microlaser array.

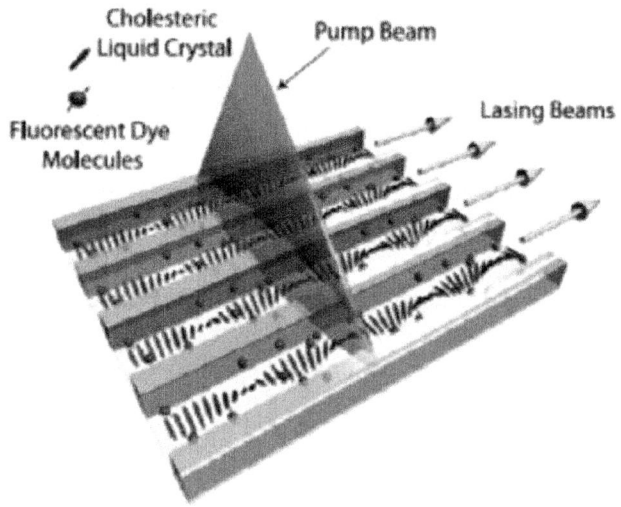

Figure 19: Sketch of a multilaser array realized in a POLICRYPS structure.

The stimulated emission emerging from the microchannels was circularly polarized, demonstrating that the distributed feedback mechanism due to the CLC helices is the cause of the observed phenomenon. The dependence of the emitted intensity and spectral linewidth (FWHM) on the input pump energy are reported in Fig. 20.

At low excitation energies, both the emission intensity and the linewidth show a quasilinear dependence on the pump energy. Above a characteristic threshold (the pump energy per excited sample area was about 5 mJ/cm^2, which corresponds to about 25 nJ/pulse), the emitted intensity suddenly starts to increase nonlinearly. Also, above this threshold, the emission linewidth breaks off from the previous trend and begins to decrease significantly. The observed pump energy value at which the power explosion and line narrowing effects occur (25 nJ/pulse) is one order of magnitude lower than in the case of other conventional dye-doped systems in a similar environment and under the same pumping conditions.

A striking scenario is presented in Fig. 21, showing the spatial distribution of the laser emission emerging from the microcavity laser array. A high sensitivity and resolution (1390 × 1024 12-bit PixelFlyQe, PCO) imaging CCD camera was employed in order to check the near-field modal profile of the stimulated emission. Images were acquired by scanning in the proximity of the output edge of our sample cell, in a direction perpendicular to the microchannels. The mapped intensity profile hereby obtained indicates that the maxima of lasing intensities have a spatial recurrence, with a periodicity that is found to be about 5 µm; this value is in perfect agreement with the initial tailoring configuration (i.e. the distance between the polymeric walls).

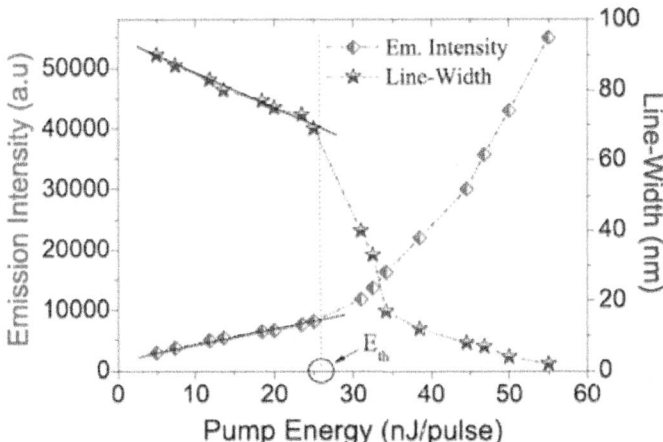

Figure 20: Emitted intensity and linewidth dependence on input pump energy. Above a threshold of 25 nJ/pulse the reported curves change from initial regimes while lasing occurs.

Therefore we can definitely conclude that POLICRYPS microchannels act as miniaturized mirrorless cavity lasers, where the emitted laser light propagates along the liquid crystal helical axis, which behaves as a Bragg resonator. This level of integration can lead to new photonic chip architectures and devices, such as a zero-threshold microlaser, phased array, discrete cavity solitons, filters, and routers. Furthermore, tailoring a proper array of electrodes, which enables the application of a local electric field, would give rise to electrically programmable phase holograms with interesting light polarization properties. Then, by including different dyes in neighbour channels, and by using proper microfibres connected at the exit, the result should be a multi-colour microlaser array with the possibility to control the intensity of each channel separately.

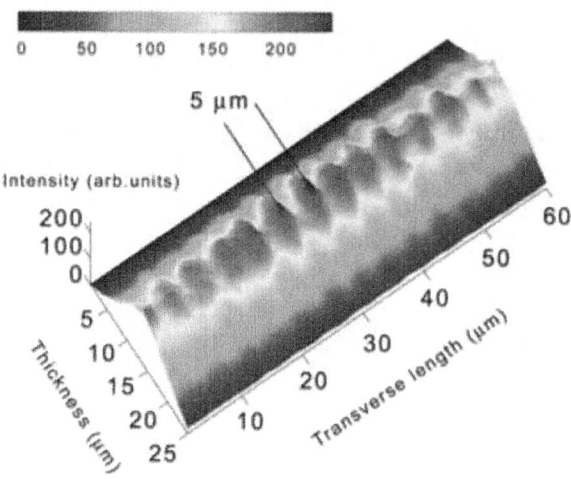

Figure 21: Spatial distribution of the laser emission emerging from the mirrorless microcavity laser array. The periodicity of maximum intensities is 5 μm. This value is in agreement with the tailoring distance between the polymeric microchannels.

POLICRYPS WITH METALLIC NANOPARTICLE INCLUSIONS

Noble metal nanoparticles (NPs) exhibiting plasmonic properties attract wide interest in research for the possibility they offer to realize metamaterials [Rockstuhl et al, 2007]. These have been predicted in 1969 by Veselago [Veselago, 1968] and they are materials that gain peculiar electromagnetic properties (e.g. negative refractive index) from their structure, rather than from their chemical composition. Thanks to recent advances in nanofabrication, first examples of such materials, which exhibit particular functionalities at optical frequencies, have been realized [Valentine et al, 2008]. However, the success of these results is limited by the typical size of devices that can be fabricated, which is actually very small (few square millimetres). Alternative approaches are emerging, which propose the use of self-assembling materials in order to overcome this issue and obtain the sought for greater structures, with less difficulty [Nanogold, (2009-2012; Metachem (2009-2013)]. An ambitious project is to combine metallic units with host materials whose dielectric properties can be tuned by an external control; indeed, a modification of the dielectric behavior of the host could correspond to a tuning action of the plasmon resonance frequency [Kossyrev et al, 2005]. In this regard, by combining the tunability of POLICRYPS structures with the plasmonic response of metallic NPs could give rise to novel metamaterial devices with tunable properties.

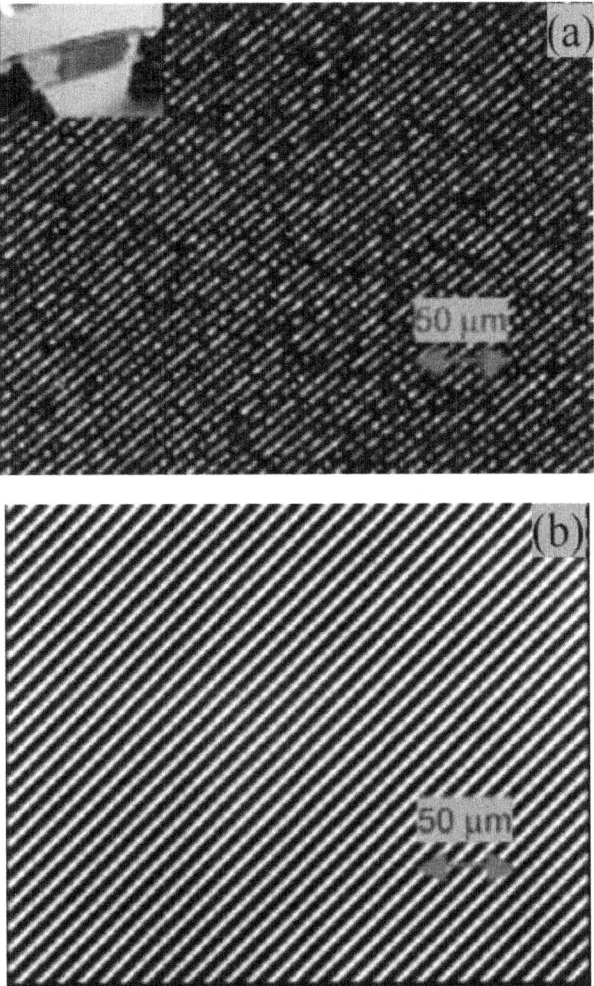

Figure 22: Polarizing optical microscope images of (a) POLICRYPS diffraction grating with Ag NP inclusions; (b) typical POLICRYPS diffraction grating. In the inset of Fig.1a, a photograph which shows the brownish color of the realized new sample

In order to obtain a POLICRYPS structure that includes metallic NPs, we have used the Harima Silver nanopaste NPS-J (from Harima Chemicals, Inc.) that is generally involved with other applications (e.g. ink-jet printing and laser sintering) [Niizeki et al, 2008]. After some difficulties encountered for mixing this Harima material with the standard POLICRYPS precursor [Caputo et al, 2011b], we obtained a new mixture composed by: NOA61, 68.5 wt%; NPS-J, 3.5 wt%; E7, 28.0 wt%. The experiment followed the same procedure used for fabricating a standard POLICRYPS structure [Caputo, 2004]: the mixture has been sandwiched by capillarity in a 13μm thick glass cell and then, by keeping it at high temperature (about 70°C), it has been exposed to a UV interference pattern with a periodicity of 6μm. A microphotograph of the fabricated sample, observed between crossed polarizers at the polarizing optical microscope (POM), is reported in Fig. 22a along with the picture of a typical POLICRYPS structure without NPs (Fig. 22b) reported for comparison aims. Some morphological differences between the two structures are evident which are obviously due to the presence of metal nano-particles in the new sample.

A SEM micrograph of the same sample (Fig. 23) reveals that the NPs are organized in clusters (the typical size ranging between 0.3μm and 1μm), homogeneously distributed all over the grating area, and visible as bright spots. Of course, those clusters that are trapped in the LC films locally disturb the order of the nematic director but, nevertheless, large LC domains, with the director aligned perpendicularly to the polymer slices of the structure, are still present. This feature, which is typical of a standard POLICRYPS grating, is evident by rotating the sample between crossed polarizers.

Figure 23: scanning electron microscope image of the POLICRYPS diffraction grating with Ag NP inclusions; Ag clusters (with size ranging between 0.3μm and 1μm) are visible in the picture as bright spots

Spectroscopical Characterization

As a consequence of considerations in paragraph 3.2, if we illuminate a POLICRYPS structure (not including metallic NPs) with linearly polarized white light (wavelength in the range 350-1100nm, at normal incidence), we expect a behaviour that is strongly dependent on the incident polarization state. In particular, we can expect that the polarization parallel to the nematic director n (p-type) is diffracted by the grating and hence the light transmission is almost suppressed. On the other hand, the orthogonal polarization (s-type) is, instead, highly transmitted in the whole analyzed range (350-1100nm), because the experienced refractive index modulation is limited and the grating is almost absent. By repeating the same experiment with the NP doped POLICRYPS sample, new features come out. As in the case of a standard POLICRYPS, the p-polarized light is diffracted by the structure and the transmission of the grating is negligible for almost all the visible part of the spectrum (Fig. 24, GPP curve). For the orthogonal polarization (s-type), we observe instead a highly transmitted intensity, whose spectrum exhibits a peculiar behaviour (Fig. 24, curve GSP). The polarization sensitivity of the grating is further on demonstrated by the GNP curve reported in the middle of Fig. 24, obtained by probing the grating with unpolarized light: the behaviour of this spectrum represents a kind of "average" of the two, differently polarized, ones. A comparison of the GSP curve (Fig. 25, top curve) with the spectrum transmitted by a mixture of Harima NPs dissolved in Chloroform (Fig. 25, bottom curve) can help to interpret above results. Indeed, the shape of this curve has particular features, exhibiting a transmission minimum at $\lambda \approx 520$nm. The typical plasmonic response of Harima NPs (20-50nm in diameter) is peaked around $\lambda = 400$nm; in presence of Ag clusters (0.3-1μm), not perfectly diluted in Chloroform, we can expect a shift of the Ag plasmonic resonance to the observed value [Mock et al, 2002]. A similar minimum, can also be noted in the GSP curve (500÷570nm) of the NP doped POLICRYPS sample. Given that the grating is almost absent for the s-polarized light (the diffraction pattern can hardly be seen in this condition), our guess is that the shape of the GSP curve of Fig. 24 reveals the presence of Ag clusters within the structure.

Above considerations can just give a qualitative proof of that guess and, as such, further investigations are essential in order to provide a quantitative confirmation. However, in case of a positive outcome, these novel structures could reveal quite promising for the realization of polarization sensitive plasmonic devices.

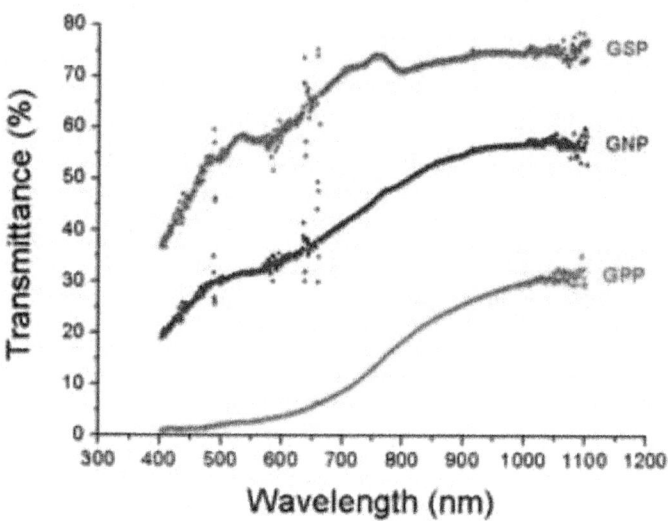

Figure 24: Spectral response of the newly realized POLICRYPS structure containing Ag NPs. Top and bottom curves have been obtained by probing the sample with s-polarized light (GSP) or p-polarized light (GPP) respectively; In the middle, the curve obtained by probing the grating with unpolarized light (GNP).

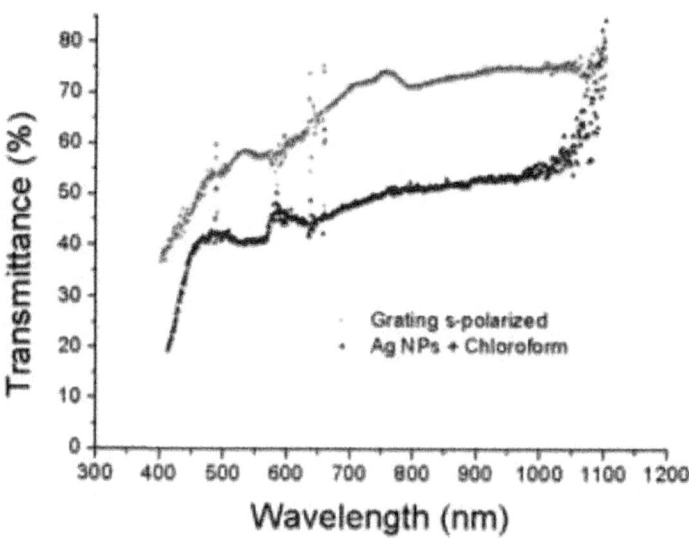

Figure 25: Comparison between the spectral response of pure Ag NPs in Chloroform (bottom curve, 5x factor) and that of Ag NPs inclusions in a POLICRYPS structure (top curve, sample probed with s-polarized white light)

CONCLUSION

In the field of electrically switchable devices, exploiting liquid crystalline composite materials, POLICRYPS represents a very promising nano/microstructure with several possibilities of application. Indeed, few main features of this system act as a common denominator for these applications: the sharpness of the structure and the uniformity of the LC films minimize light scattering losses, while the application of a suitable, relatively low, external voltage can determine, in a millisecond timescale, a reorientation of the LC director and hence the tunability of the device. Depending on the way a light beam propagates through, the POLICRYPS can be used as a switchable diffraction phase grating, for light impinging at a given angle with the structure; a switchable optical phase modulator (with a light beam impinging almost perpendicularly to the structure); an all-optical switchable device, an array of mirrorless optical micro-resonators devoted to obtain a tuneable lasing effect (if the NLC is substituted with a mixture of dye-doped CLC and the system is optically pumped) and a plasmonic device with polarization sensitive properties (when metallic nanoparticles are included in the chemical mixture). Performances exhibited in all the above applications are very interesting and stimulate further investigations in the different fields.

ACKNOWLEDGMENT

Our sincere thanks go to Dr. Giuseppe Strangi and his group for the fruitful collaboration in the realization and characterization of microlaser arrays in POLICRYPS structures and to Dr. Nelson Tabiryan and his group for the possibility offered in realizing Azo-materials doped POLICRYPS structures. Finally, the research leading to these results has received funding from the European Union's Seven Framework Programme (FP7/2007-2013) under grant agreement n°228455.

REFERENCES

1. Bunning, T.J.; Natarajan, L.V.; Tondiglia, V.P. & Sutherland, R.L. (2000). Holographic polymerdispersed liquid crystals (H-PDLCs), Annu. Rev. Mater. Sci. Vol. 30, pp. 83-115

2. Caputo, R.; Sukhov, A.V.; Umeton, C.P. & Ushakov, R.F. (2000). Formation of a grating of submicron nematic layers by photopolymerization of nematic-containing mixture, J. Exp. Theor. Phys. Vol. 91 pp. 1190–1197

3. Caputo, R.; Sukhov, A.V.; Tabyrian, N.V.; Umeton, C.P. & Ushakov, R.F. (2001). Mass transfer processes induced by inhomogeneous photo-

polymerisation in a multicomponent medium, Chem. Phys. Vol. 271, pp. 323–335

4. Caputo, R.; De Sio, L.; Sukhov, A.V.; Veltri, A. & Umeton, C.P. (2004). Development of a new kind of switchable holographic grating made of liquid crystal films separated by slices of polymeric material (policryps), Opt. Lett. Vol. 29, pp. 1261-1263

5. Caputo, R.; Umeton, C.P.; Veltri, A.; Sukhov, A.V. & Tabiryan, N. (2007). Holographic diffraction grating,process for its preparation and opto-electronic devices incorporating it, European Patent Request 1649318; US Patent Request 2007/0019152A1

6. Caputo, R.; Trebisacce, I.; De Sio, L. & Umeton, C.P. (2010). Jones matrix analysis of dichroic phase retarders realized in soft matter composite materials, Opt. Express Vol. 18, pp. 5776-5784

7. Caputo, R.; Trebisacce, I.; De Sio, L. & Umeton, C.P. (2011a). Phase modulator behavior of a wedge-shaped POLICRYPS diffraction grating, Mol. Cryst. Liq. Cryst. (accepted)

8. Caputo, R.; De Sio, L.; Dintinger, J.; Sellame, H.; Scharf, T. & Umeton, C.P. (2011b).Realization and characterization of POLICRYPS-like structures including metallic subentities, Mol. Cryst. Liq. Cryst. (submitted)

9. de Gennes, P.G. (1993). The Physics of Liquid Crystals (Oxford: Clarendon)

10. De Sio, L.; Caputo, R.; De Luca, A.; Veltri, A.; Sukhov, A.V. & Umeton, C.P. (2006). In situ optical control and stabilization of the curing process of policryps gratings, Appl. Opt. Vol. 45, pp. 3721-3727

11. De Sio, L.; Veltri, A.; Tedesco, A.; Caputo, R.; Sukhov, A.V. & Umeton C.P. (2008a). Characterization of an active control system for holographic set-up stabilization, Appl. Opt. Vol. 47, pp. 1363-1367

12. De Sio, L.; Tabiryan, N.; Caputo, R.; Veltri, A. & Umeton C.P. (2008b). Policryps structures as switchable optical phase modulators, Opt. Express Vol. 16, pp. 7619–7624

13. De Sio, L.; Serak, S.; Tabiryan, N.; Ferjani, S.; Veltri, A. & C. Umeton, (2010). Holographic Gratings Containing Light-Responsive Liquid Crystals for Visible Bichromatic Switching, Adv. Mater. Vol. 22, pp. 2316-2319.

14. EU Project: "Self-organised nanomaterials for tailored optical and electrical properties (Nanogold)", FP7-NMP-SMALL-2008-228455, nanogold.epfl.ch, (2009-2012);

15. EUProject: "Nanochemistry and self-assembly routes to metamaterials for visible light (Metachem)", FP7-NMP-SMALL-2009-228762, www. metachem-fp7.eu, (2009-2013).

16. Goldberg, L.S. & Schnur, J.M. (1973). Tunable internal-feedback liquid crystal laser, US Patent Specification 3 771 065

17. Kogelnik, H. (1969). Coupled Wave Theory for Thick Hologram Gratings, Bell Syst. Tech. J. Vol. 48, pp. 2909-2948

18. Kogelnik, H. & Shank, C.V. (1971). Stimulated emission in a periodic structure, Appl. Phys. Lett. Vol. 18, pp. 152-154

19. Kossyrev, P.A.; Yin, A.; Cloutier, S.G., Cardimona, D.A.; Huang, D.; Alsing, P.M. & Xu, J.M.; (2005) Electric Field Tuning of Plasmonic Response of Nanodot Array in Liquid Crystal Matrix, Nano. Lett. Vol. 5, pp. 1978-1981.

20. Lucchetta, D.E.; Criante, L. & Simoni, F. (2003) Optical characterization of polymer dispersed liquid crystals for holographic recording, J. Appl. Phys. Vol. 93, pp. 9669-9675

21. Margerum J.D.; Lackner, A.M.; Ramos, E.; Smith, G.W.; Vaz, N.A.; Kohler, J.L. & Allison, C.R. (1992). Polymer dispersed liquid crystal film devices, US Patent Specification 5,096,282, March 17, 1992

22. Mock, J.J.; Barbic, M.; Smith, D.R.; Schultz, D.A. & Schultz, S. (2002). Shape effects in plasmon resonance of individual colloidal silver nanoparticles, J. Chem. Phys. Vol. 116, 6755-6759

23. Niizeki, T.; Maekawa, K.; Mita, M.; Yamasaki, K.; Matsuba, Y.; Terada, N. & Saito, H.; (2008) Laser Sintering of Ag Nanopaste Film and Its Application to Bond-Pad Formation, Proc. ECTC 2008.

24. Ozaki, M.; Kasano, M.; Ganzke, D.; Haase, W. & Yoshino, K. (2002) Mirrorless lasing in a dyedoped ferroelectric liquid crystal, Adv. Mater. Vol. 14, pp. 306-309

25. Rockstuhl, C.; Lederer, F.; Etrich, C.; Pertsch, T & Scharf, T.; (2007). Design of an Artificial Three-Dimensional Composite Metamaterial with Magnetic Resonances in the Visible Range of the Electromagnetic Spectrum, Phys. Rev. Lett. Vol. 99, 017401.

26. Strangi, G.; Barna, V.; Caputo, R.; De Luca, A.; Versace, C.; Scaramuzza, N.; Umeton, C.P. & Bartolino, R. (2005) Color-tunable organic microcavity laser array using distributed feedback, Phys. Rev. Lett. Vol. 94, 063903

27. Sutherland, R.L.; Tondiglia, V.P.; Natarajan, L.V.; Bunning, T.J. & Adams W.W. (1994).Electrically switchable volume gratings in polymer-dispersed liquid crystals, Appl. Phys.Lett., Vol. 64, pp. 1074–1076

28. Sutherland R.L., Tondiglia V.P., Natarajan L.V., Bunning T.J. & Adams W.W. (1996). Electrooptical switching characteristics of volume holograms in polymer dispersed liquid crystals,J. Nonlinear Opt. Phys. Mater., Vol. 5, pp. 89–98

29. Tsutsumi, O. & Ikeda, T.; (1995). Optical switching and image storage by means of azobenzene liquid-crystal films, Science Vol. 268, pp. 1873-1875

30. Yariv, A. (1989). Quantum Electronics (Wiley, New York) Valentine, J.; Zhang, S.; Zentgraf, T.; Ulin-Avila, E.; Genov, D.A.; Bartal, G. & Zhang, X.; (2008). Three Dimensional Optical Metamaterial Exhibiting Negative Refractive Index,Nature, vol. 455, 376.

31. Veselago, V.G.; (1968). The electrodynamics of substances with simultaneously negative values of ε and μ, Sov. Phys. Usp, Vol. 10, 4, 509-514 (1968).

32. Wu, Y.H.; Lin, Y.H.; Lu, Y.Q.; Ren, H.; Fan, Y.H.; Wu, J. & Wu, S.T. (2004). Submillisecond response variable optical attenuator based on sheared polymer network liquid crystal, Opt. Express Vol. 12, pp. 6382–6389

Chapter 5

DIGITAL MICROSCOPY AND IMAGE ANALYSIS APPLIED TO COMPOSITE MATERIALS CHARACTERIZATION

S. Paciornik; J. d'Almeida
DEMa PUC-Rio, Rio de Janeiro, Brazil

ABSTRACT

Digital Microscopy was employed to characterize the microstructure of fiber-reinforced composite tubes manufactured by filament winding. Optical Microscopy was used for void characterization while Scanning Electron Microscopy was used for fiber and layer analysis. Acquired images were assembled in mosaics to reveal the microstructure of different cross-sections of the sample. Image processing was employed to detect either voids or individual fibers and measure their size, shape and spatial distribution. Void spatial distribution was analyzed with two different methods - local analysis and the tessellation method - revealing different behaviors along different cross-sections. Fiber layers were automatically detected and their average winding angle and dispersion were analyzed.

INTRODUCTION

The microstructural characterization of a material is a key step to understand its practical engineering properties. For composites, the microstructural characterization is even more critical, because of the many possible arrangements between the reinforcing phase and the matrix. When one wants to characterize a composite, the volume or mass fraction of the phases is the most commonly described parameter. This is a direct consequence of the strong influence that the volume fraction of the reinforcing phase has on the properties of the composite.

However, for heterogeneous materials like fiber-reinforced composites, the influence of the microstructure on the engineering properties is critical, and besides the volume fraction, the complete understanding of composites properties requires the determination of several microstructural parameters such as size, orientation and spatial distribution of the fibers and of possible defects. In fact, voids are a common feature appearing during the usual manufacturing processes [1, 2, 3], and even for aeronautical grade resin matrix composites, the presence of voids is allowed although restricted to low percentages. But, voids are stress concentrators and can act as crack initiation points [4]. Therefore, besides their volume fraction, other microstructural characteristics such as their spatial distribution and the aspect ratio should be carefully evaluated [1].

Although measuring the volume fraction of voids or particles is a common practice [5, 6], the spatial distribution of these microstructural parameters can also be of relevance [1]. By characterizing the preferred distribution of voids, one can access information about the manufacturing process itself. For example, for filament winding parts, voids preferentially distributed among the tows of fibers are probably due to an inappropriate level of stress when the fibers are being wound around the mandrel [7]. On the other hand, voids on resin rich areas can originate from entrapped air bubbles generated during the stirring of the resin or are due to the evolution of by-products during the cure of the resin. Moreover, it can be of interest to determine if the voids are clustered or uniformly distributed on the cross section of the composite.

The use of filament winding also brings a new level of complexity to the characterization of the reinforcement phase, as the composite is made up of a series of layers, with different thickness and fiber orientations. So it is important to discriminate individual fibers, their orientations, and groups of similarly oriented fibers forming layers. Gathering all this information is, however, not easy using the traditional methods of microstructural analysis.

Digital Microscopy (DM) is the convergence of microscope automation, digital image acquisition, processing and analysis. The use of DM allows a complete characterization of a sample in fully automated procedures. Online acquired images can be automatically treated and the desired features can be discriminated and analyzed. Moreover, a large number of fields, comprising thousands of objects can be quickly analyzed, providing high statistical accuracy [8].

In this work, the microstructure of a glass fiber-reinforced composite pipe fabricated by filament winding was fully characterized by Digital Microscopy. Both Optical Microscopy (OM) and Scanning Electron Microscopy (SEM) were employed, respectively, for void and fiber characterization.Materiais e

EXPERIMENTAL

Specimen Selection and Preparation

Polyester matrix-filament wound glass fiber pipes used for the transport of water in offshore oil production facilities and having a nominal internal diameter of 200 mm and 7 mm thickness, were analyzed in this work. The pipes were sectioned in rings 25 mm thick that were then cut into 4 quadrants. Axial and circumferential samples were prepared for microstructural analysis (Figure 1). The preparation followed the usual procedures of grinding and polishing, from silicon carbide grit (#100) to alumina powder (0.5 μm). Whenever possible the cutting operation was performed using a low speed saw, to avoid excessive damage of the glass fibers [9].

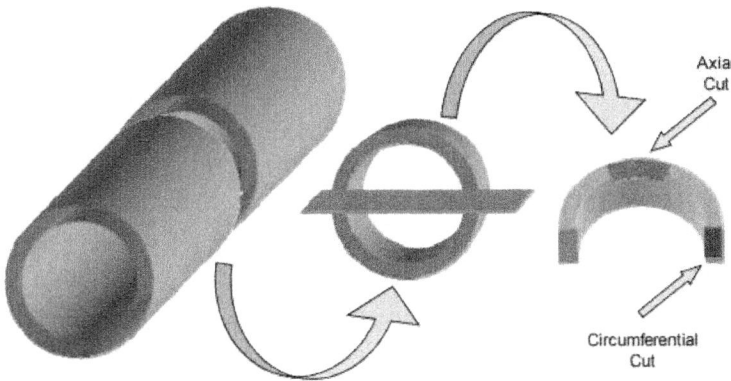

Axial
Cut

Circumferential
Cut

Figure 1: Illustration of axial and circumferential samples obtained from the composite tube.

Image Acquisition

To characterize the microstructure in respect to the distribution of voids and fibers, as well as their spatial orientation, it was desirable to obtain images combining high magnification and a large area of analysis. These requisites are not fulfilled by a single image. Therefore, it was necessary to capture several images, with the same magnification, and to join them together generating a mosaic image [10].

For void characterization, the samples were observed by optical microscopy using a Zeiss AxioPlan 2 motorized optical microscope. The use of this computer-controlled equipment, with a motorized x-y-z sample holder, allows controlled sample displacement and the acquisition of a sequence of images with any spatial distribution and with automated focus control [8]. The images were captured using an AxioCam HR digital camera, with 1300 x 1300 pixels resolution.

Image mosaics were constructed by joining low magnification images, obtained using a 5x objective lens (Zeiss EPIPLAN, NA = 0.13). These images covered an entire cross-section of the samples, from the outer to the inner diameter. Each field occupied an area of 2750 x 2180 μm^2, with a spatial resolution of 2.10 μm/pixel. To cover the entire sample area, 7 fields on the x direction and 3 on the y direction were necessary, leading to a total area of 19.2 x 6.5 mm^2. This procedure permitted a complete visualization of each sample, clearly revealing the spatial distribution of voids, and allowing the measurement of features on the entire sample.

For fiber and layer characterization the images the samples were observed by scanning electron microscopy, using the backscattered (BSE) mode of image formation. BSE was preferred instead of secondary electrons imaging, because the intensity of the BSE signal is a function of the atomic weight of the elements on the sample. Therefore, a good contrast is obtained between the polymeric matrix and glass fibers. Moreover, the images do not show topographical information, which is beneficial in the image processing steps, as residual artifacts from sample preparation are blurred. The images were captured with 512 x 480 pixels resolution, at a magnification of 200x, corresponding to 0.88 μm/pixel.

In this case, mosaic images were generated combining 28 images - 4 fields in the x direction and 7 in the y direction. This procedure allowed for a complete visualization of the cross-section, clearly revealing the spatial distribution of the fiber layers, and allowing the measurement of features on the entire sample.

RESULTS AND DISCUSSION

Spatial Distribution of Voids

Figure 2a shows the image of a mosaic from a circumferential sample. The larger white rectangle depicted represents one of the 21 (7 x 3) mosaic tiles, also shown magnified in Figure 2b. As one can see, the mosaic offers a global view of the sample, clearly showing the spatial distribution of voids and fibers. In this view, the fibers are mainly aligned perpendicular to the plane of cut and appear as circles. Most large visible voids are also approximately round, as can be seen in Figure 2b. It is worth mentioning that the circumferential sample actually shows the spatial distribution of voids along the axis of the tube.

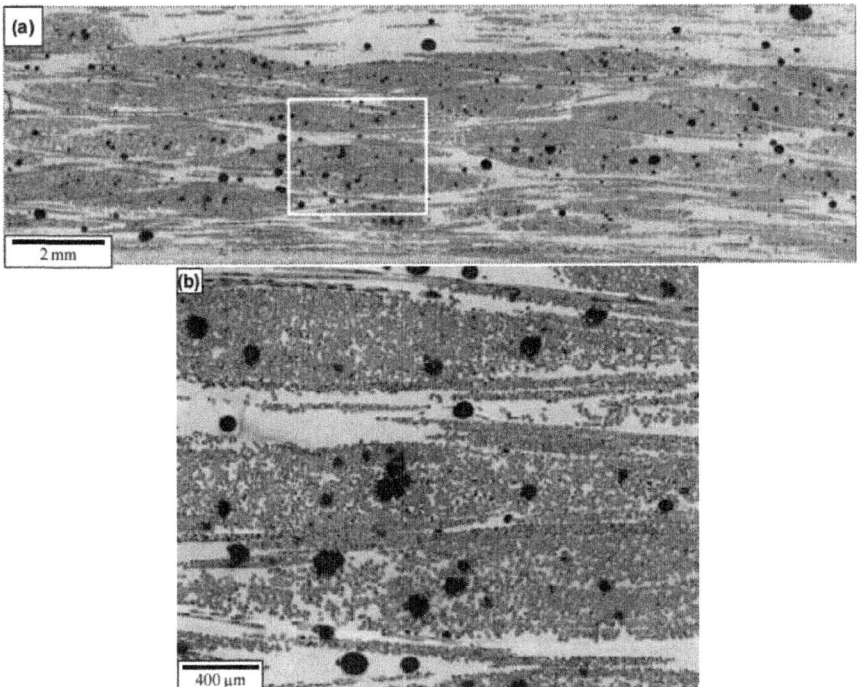

Figure 2: Circumferential sample. (a) Mosaic image obtained with a 5X objective lens (2.1 μm/pixel). (b) Digital magnification of the area outlined by the white rectangle.

Figure 3a shows a mosaic image of an axial cut through the sample. The fibers appear as elongated objects while both round and elongated voids are visible. The presence of very elongated voids (Figure 3b), illustrates the relevance of observing the sample in different orientations. Moreover, such long objects also highlight the usefulness of the mosaic image - in a regular field scan microscopy procedure these objects can be too large to fit any individual field, precluding their accurate characterization of size and shape. Again, it is worth mentioning that the axial sample actually shows the spatial distribution of voids along the circumference of the tube.

Voids can be discriminated by their grey shade. As seen in Figures 2 and 3, they appear darker than the fibers and the polymer matrix. However, small regions of fibers damaged during sample preparation exhibit similar contrast, and may lead to wrong results. Therefore, an image processing and analysis routine was developed to discriminate the several kinds of dark regions appearing on the images, and recognize the voids. This routine is described in detail elsewhere [10]. The results are shown in Figure 4 for a representative field in Figure 2a.

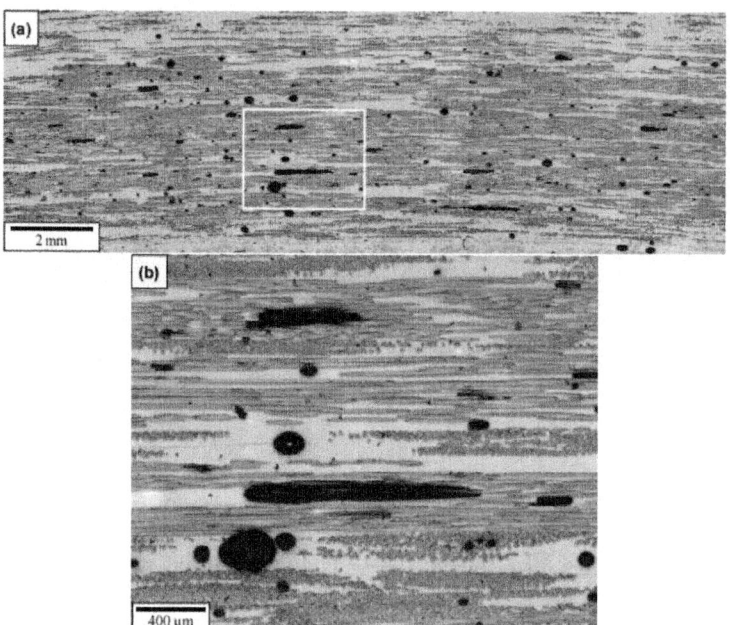

Figure 3: Axial sample. (a) Mosaic image obtained with a 5X objective lens (2.1 μm/ pixel). (b) Digital magnification of the area outlined by the white rectangle.

Once voids are reliably recognized, their spatial distribution can be measured. This analysis was performed on the mosaic images, with two different methods: local mapping and neighborhood analysis. In local mapping the mosaic images were scanned with an analysis window of 250 x 250 pixels and, for each window, two parameters were obtained: void count and void area fraction. To avoid multiple counting of objects at the edges of the windows, a guard frame was used in which voids touching the bottom or left edges of the window were not considered. Thus, if a void straddles an edge between windows it will be counted only once. These parameters were then plotted as contour maps as shown in Figures 5 and 6 for the circumferential and axial samples, corresponding to Figures 2a and Figure 3a.

The maps provide a global view of the desired parameters across the whole cross-section, and indicate variations in spatial distribution. Comparing Figure 5 to Figure 6 one can state that the spatial distribution of voids for the circumferential sample is more uniform than for the axial sample. However, even though the maps contain quantitative information, the level of uniformity of the spatial distribution is still deduced in a qualitative way. To obtain a quantitative parameter that describes the uniformity of spatial distribution, a neighborhood analysis was applied.

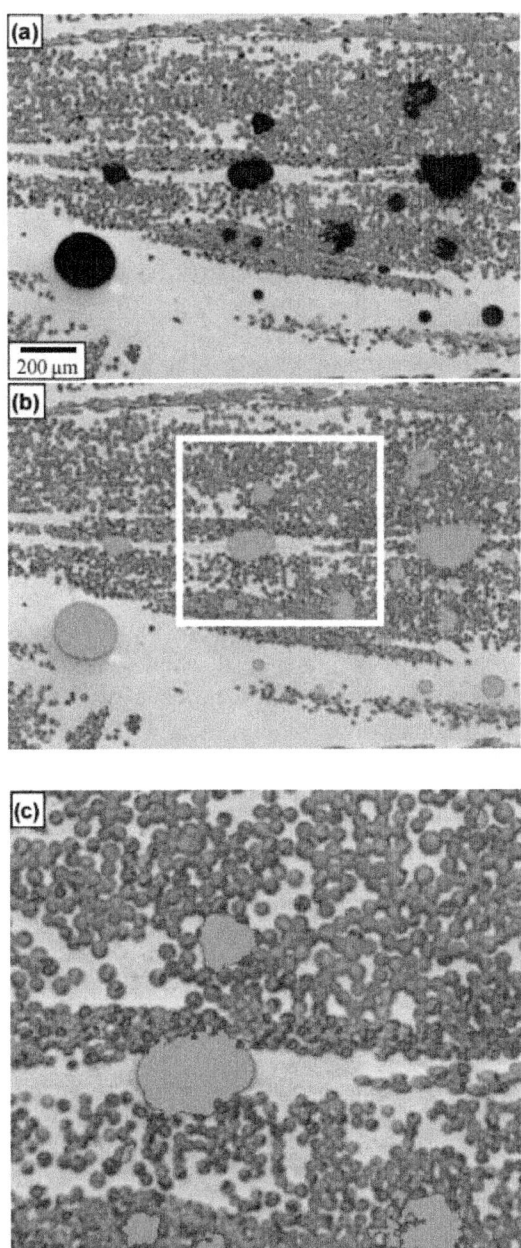

Figure 4: Void discrimination. (a) Original image (5X objective lens, 2.1 μm/pixel). (b) Detected dark objects (green and red). (c) Magnified view of the region outlined in (b) – voids are shown in green while red objects are smaller than 40 pixels in area and correspond to broken fiber tips.

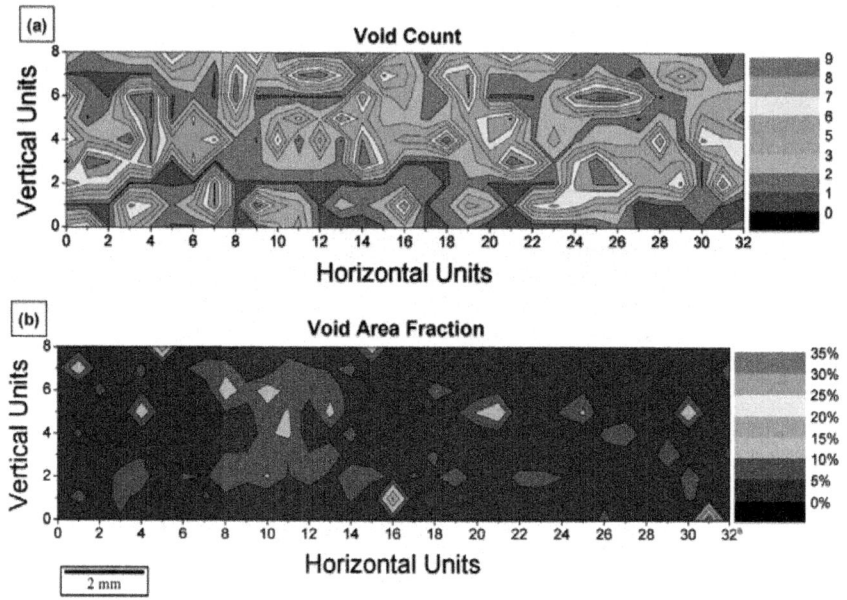

Figure 5: Local mapping for circumferential sample. (a) Void count. (b) Void area fraction. Horizontal and vertical units refer to the count of 250x250 pixels windows in each direction. See text for details.

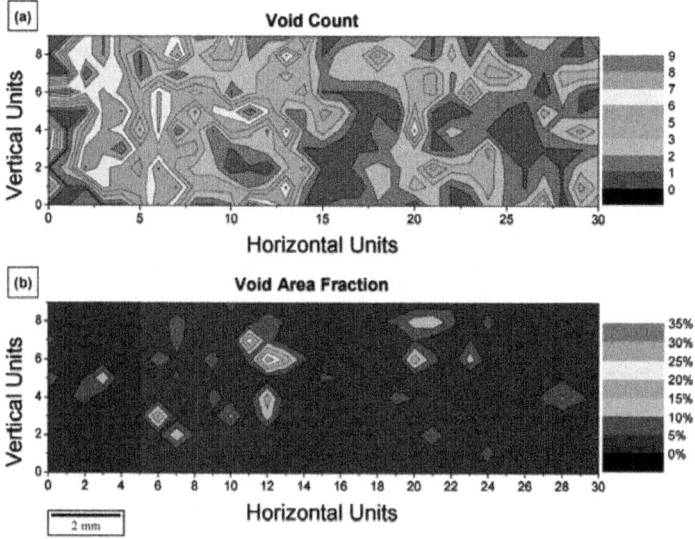

Figure 6: Local mapping for axial sample. (a) Void count. (b) Void area fraction. Horizontal and vertical units refer to the count of 250x250 pixels windows in each direction. See text for details.

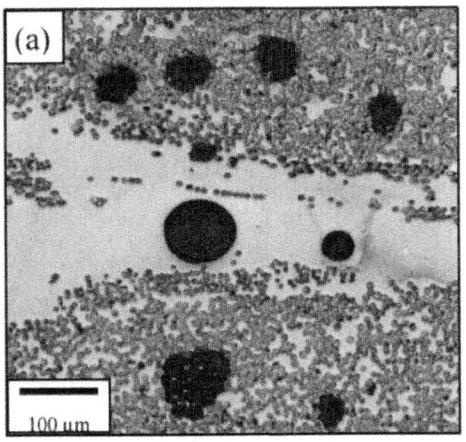

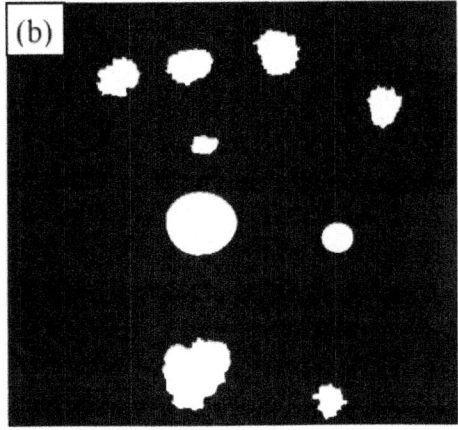

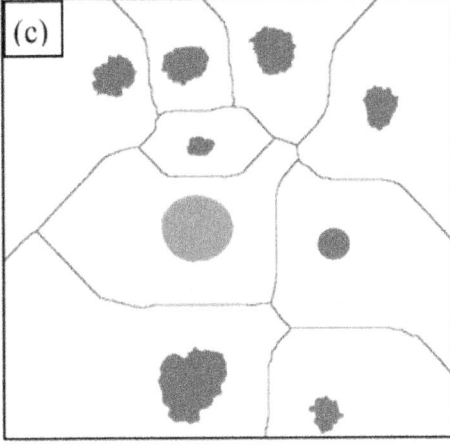

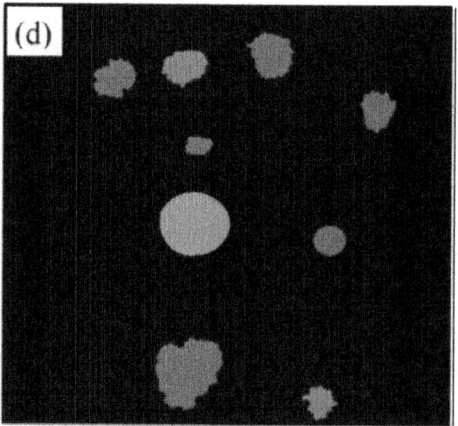

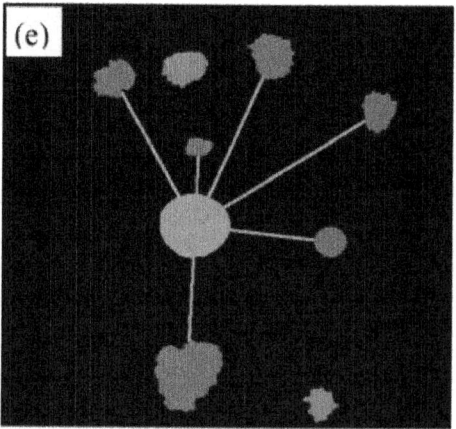

Figure 7: Image processing sequence to measure void nearest neighbors' distances. (a) Original image fragment. (b) Detected voids. (c) Voids (in color) superimposed on their regions of influence. The reference void is shown in green. (d) Detection of nearest neighbors (shown in red). Grey voids are not nearest neighbors. (e) Determination of distances between reference void and its neighbors. (f) Distribution of distances.

In this analysis, the neighborhood of each relevant object, a void in this case, is analyzed and its nearest neighbors are identified by the so-called tessellation technique [11]. See Figure 7 for the sequence steps shown in a small section of one of the mosaics. The original grayscale image (Figure 7a) is segmented and post-processed as described above, leading to a binary image showing the voids (Figure 7b). All particles are then simultaneously dilated until a one pixel boundary is left around each particle (Figure 7c). This technique is based on the well-known Voronoi diagram [12]. Through a sequence of dilation and intersection, the neighboring regions to any given reference region can be automatically identified. Another intersection operation shows the corresponding original voids (Figure 7d). Then, the edge to edge distances between the reference void and its nearest neighbors are obtained (Figure 7e). A set of vectors is obtained for each void (Figure 7f) and the sequence is repeated for each and every void.

Thus, a distribution of distances is measured for each void, with its average and standard deviation. The coefficient of variation (cov) is defined as the ratio of the standard deviation to the average distance. It has been shown previously [11] that for a uniform random spatial distribution $cov = 0.36 \pm 0.02$. Deviations from this value indicate some degree of clustering of the analyzed phase.

For the circumferential and axial samples shown, respectively, in Figures 2 and 3, the values obtained were $cov = 0.37$ and $cov = 0.50$. These results confirm the qualitative analysis that indicated that the spatial distribution of voids was almost perfectly random for the circumferential sample while the axial sample showed substantial clustering.

Fiber and Layer Analysis

Figure 8a shows a typical mosaic obtained operating the SEM in BSE mode. The contrast between fibers and matrix is sharp and one can easily distinguish fibers in different orientations, and the several layers resulting from filament winding. This image has undergone background correction to eliminate intensity variations between the several image tiles and noise filtering. The details of the procedure are described elsewhere [13].

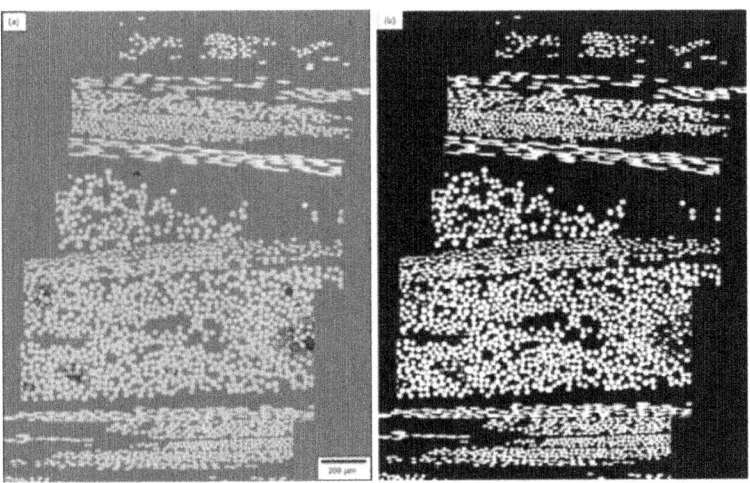

Figure 8: (a) Example of a mosaic image obtained by assembling 28 (4x7) individual images, after equalizing the illumination of each field. (b) Fiber discrimination.

Fibers were automatically discriminated by their gray shade, as shown in Figure 8b. A few steps of post-processing were necessary to eliminate spurious objects in the background and holes in fibers. Boundaries between touching fibers were also obtained by the traditional watershed method [14].

Once fibers were accurately detected, a sequence of image processing operations was employed to discriminate layers. The basic principle behind this discrimination was the separation by fiber shape. Fibers normal to the image plane appear round, while fibers inclined relative to this plane appear as ellipsoids. Thus, shape parameters such as the aspect ratio and the circular shape factor [8] where used. Thus, the original image was separated into two intermediate images containing normal and inclined fibers. See Figure 9.

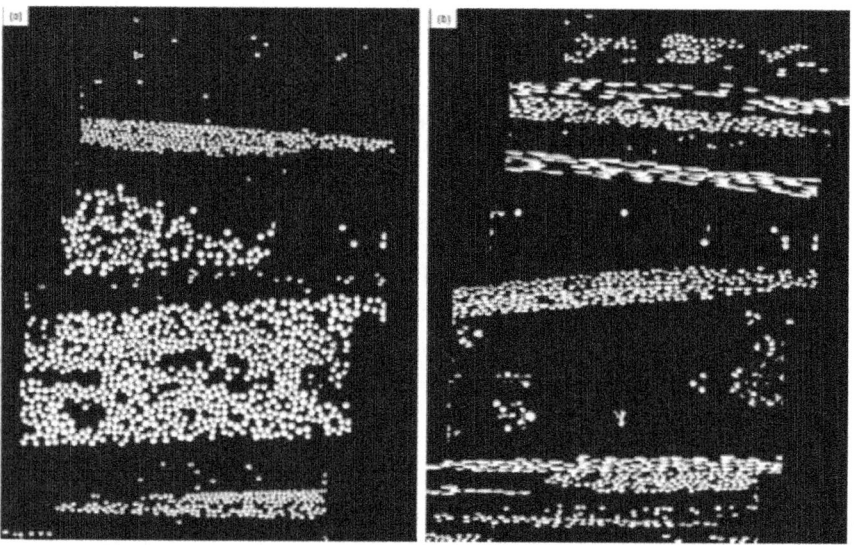

Figure 9: (a) Fibers normal to the image plane. (b) Fibers inclined to the image plane.

A sequence of morphological operations was then applied to join neighboring fibers, eliminate stray or broken fibers, and detect each fiber layer. The details are described elsewhere [13]. The result is shown in Figure 10, where each layer is identified by a different color, superimposed on the original image.

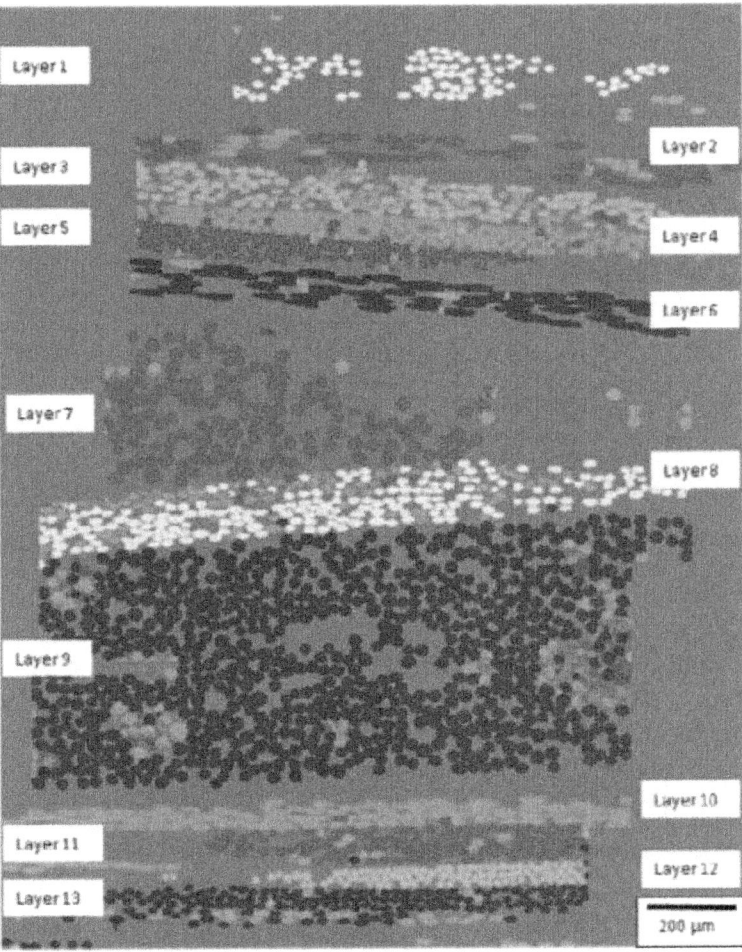

Figure 10: Final discrimination of fiber layers according to fiber orientation.

The winding angle of each fiber can be obtained through the measurement of its major and minor projections (F_{max} and F_{min}) and the equation

$$\phi_{Fiber} = arcsen\left(F_{min}\middle/F_{max}\right) \cdot \left(\frac{180}{\pi}\right)$$

(1)

the winding angle of a layer is estimated as the average of the winding angles of all fibers that belong to that layer. The results are show in Table 1, where one can see that a nonsymmetrical stacking sequence was used. From a mechanical behavior point of view this is not a good result, since nonsymmetrical laminates have bending-stretching coupling that can cause warping due to in-plane forces [15].

Table 1: Winding angles of the fiber layers

Layer #	Angle (degrees)*
1	38.8 ± 4.9
2	14.0 ± 3.8
3	34.4 ± 11.4
4	49.4 ± 5.0
5	63.6 ± 5.3
6	10.5 ± 3.2
7	63.0 ± 4.6
8	36.6 ± 10.5
9	63.2 ± 4.0
10	7.5 ± 4.0
11	36.9 ± 10.1
12	64.9 ± 2.8
13	41.2 ± 8.8

It is apparent from the results listed in Table 1 that larger standard deviations were found for the layers with more inclined angles (especially layers 3, 8, and 11). This is consistent with the comments above, regarding the difficulty to separate touching fibers in these layers. In certain cases, the separation lines of the watershed method break elongated fibers, creating objects with varying shapes within a layer, and thus biasing the winding angle measurement.

The methodology of identification of the actual lamina stacking sequence, however, proves its robustness, since the microstructural characterization of the entire cross section can be used to monitor deviations from the manufacturing designed lamina stacking sequence. A correction in the fabrication routine can thus be made, and better quality pipes can be manufactured.

CONCLUSIONS

The use of mosaic images covering the entire cross section of the fiber-reinforced composite was fundamental to reveal the complete microstructural arrangement, showing the spatial distribution of voids and fibers.

The developed methodology was able to discriminate voids from preparation defects, and obtain their spatial distribution in a quantitative way. It was also possible to identify each individual lamina wound during the manufacturing process of a composite pipe. The image processing and analysis procedure developed to measure fiber size and shape allowed the determination of the average winding angle for each identified layer.

ACKNOWLEDGMENTS

The support of CNPq is gratefully acknowledged.

REFERENCES

1. HAMIDI, Y.K., ALTAN, M.C., "Spatial variation of void morphology in resin transfer molded E-glass/epoxy composites", *Journal of Materials Science Letters*, v. 22, pp. 1813-1816, 2003.

2. SANTULLI, C., GIL, R.G., LONG, A.C., CLIFFORD, M.J., "Void content measurements in commingled E-glass/polypropylene composites using image analysis from optical micrographs", *Science Engineering Composites Materials*, v. 10, pp. 77-90, 2002.

3. SAINT-MARTIN, G., SCHMIDT, F., DEVOS, P., LEVAILLANT, C., "Voids in short fibre-reinforced injection-molded parts: density control vs. mass control", *Polymer Testing*, v. 22, pp. 947-953, 2003.

4. BAI, J., HU, G., BOMPARD, P., "Mechanical behaviour of ± 55o filament-wound glass-fibre/epoxy-resin tubes", *Composites Science and Technology*, v. 57, pp. 155-164, 1997.

5. CILLEY, E., ROYLANCE, D., SCHNEIDER, N., "Methods of fiber and void measurement in graphite/epoxy composites", *In Composite Materials: Testing and Design (Third Conference), ASTM STP 546*, American Society for Testing and Materials, pp. 237-249, 1974.

6. PURSLOW, D., "On the optical assessment of the void content in composite materials", *Composites*, v. 15, pp. 207-210, 1984.

7. MCCARVILL, W.T., "Filament-winding resins", In: Engineered Materials Handbook - vol.1", Composites, ASM International, Metals Park, pp. 135-138, 1987.

8. PACIORNIK, S., MAURICIO, M.H.P., "Digital imaging", *In: ASM Handbook: Metallography and Microstructures, ASM International, Materials Park*, pp. 368-402, 2004.

9. VOORT, G.F.V., "Metallography, principles and practice", *ASM International, Materials Park*, 2000.

10. PACIORNIK, S., D'ALMEIDA, J.R.M., "Measurement of void content and distribution in composite materials through digital microscopy", *Journal of Composite Materials*, v. 43, n. 2, pp. 101-112, 2009.

11. YANG, N., BOSELLI, J., SINCLAIR, I., "Simulation and quantitative assessment of homogeneous and inhomogeneous particle distributions in

particulate metal matrix composites", *Journal of Microscopy*, v. 201, pp. 189-200, 2000.

12. SHEHATA, M.T., "Characterization of particle dispersion", *In: Practical Guide to Image Analysis. ASM International, Materials Park*, pp. 129-144, 2000.

13. OLIVEIRA, J.G.A., PACIORNIK, S., D'ALMEIDA, J.R.M., "Microstructural analysis of composite yubes through digital microscopy", *Journal of Composite Materials*, v. 43, pp. 1857-1868, 2009.

14. BEUCHER, S., LANTEJOUL, C., "Use of watersheds in contour detection", *In: Proceedings of the International Workshop on Image Processing, Real-time Edge and Motion Detection/Estimation*, 2.1-2.12, Rennes, France, 1979.

15. GIBSON, R.F., *Principles of composite material mechanics*, McGraw-Hill, NewYork,NY, 1994.

Chapter 6

EFFECT OF INTERFACE STRUCTURE ON MECHANICAL PROPERTIES OF ADVANCED COMPOSITE MATERIALS

Yong X. Gan

Department of Mechanical, Industrial and Manufacturing Engineering, College of Engineering, University of Toledo, 2801 W Bancroft Street, Toledo, OH 43606, USA

ABSTRACT

This paper deals with the effect of interface structures on the mechanical properties of fiber reinforced composite materials. First, the background of research, development and applications on hybrid composite materials is introduced. Second, metal/polymer composite bonded structures are discussed. Then, the rationale is given for nanostructuring the interface in composite materials and structures by introducing nanoscale features such as nanopores and nanofibers. The effects of modifying matrices and nano-architecturing interfaces on the mechanical properties of nanocomposite materials are examined. A nonlinear damage model for characterizing the deformation behavior of polymeric nanocomposites is presented and the application of this model to carbon nanotube-reinforced and reactive graphite nanotube-reinforced epoxy composite materials is shown.

INTRODUCTION

Advanced composite materials have the unique combination of outstanding mechanical properties of matrices and reinforcements. The reinforcement/matrix interface in composite materials forms in manufacturing processes and determines the performances of the composite materials. Some reinforcements may not be compatible with matrices in view of their physical and/or chemical properties, which causes premature failure of the composites. For example, ultrahigh molecular weight polyethylene (UHMWPE) fibers have poor wettability with epoxies. As a result, the interface bonding strength between the fibers and polymer matrices is very low. Recently, development of nanofiber modified matrices containing reactive graphitic nanofibers (r-GNFs) has been proposed to promote the wetting of the matrices to certain types of fiber reinforcements. In this paper, the effect of interface structures on the mechanical properties of fiber reinforced composite materials is discussed. The wettability of UHMWPE fibers with different epoxy matrices including a surface modified carbon nanotube-containing epoxy and a pure epoxy is presented. How to change the interface structures to reduce the contact angle between the epoxies and the fibers is described. Finally, the nonlinear damage model for evaluating the mechanical property change associated with the interface damage is presented.

HYBRID COMPOSITE MATERIALS

Composite materials are designed to have a combination of the properties of each of the components. There are several types of composite materials, including particle-reinforced, fiber-reinforced composite materials, *etc*. Hybrid composite materials containing continuous fiber reinforced plies and metal layers are special composite materials because of their high specific strength, high specific modulus, excellent electromagnetic shielding characteristics and very good high-cycle fatigue property. Typical examples of such composites include carbon fiber reinforced aluminum foams or laminates (CARE) [1], Kevlar fiber reinforced aluminum hybrid composites (ARALLs) [2]. These composite materials consist of alternating layers of metal sheets and fiber reinforced epoxy composites. The unique properties of the fiber reinforced epoxy composites are retained and the materials are immune to environmental attack due to the incorporation of the sandwiched metal layers. The metal layers are also responsible for providing high shear strength. Therefore, applications of such hybrid composite materials in aerospace, electronics and automotive industries have been considered [3].

Hybrid composite materials/structures are frequently subjected to thermal and mechanical fatigue loading. Aside from external mechanical loadings, thermal effect is identified as an important factor that determines the stress distribution in composite materials. During the curing process, adhesively bonded composite/metal laminate structures are held at elevated temperatures over 120 °C, very high residual stresses could build up because of the difference in coefficients of thermal expansion (CTE) for different materials. The CTE of aluminum is about $2.36 \times 10^{-5}/°C$ and for polymers it is higher than $1.05 \times 10^{-4}/°C$ [4]. This thermal mismatch results in delamination or debonding of hybrid composite materials, which facilitates fatigue crack growth in the polymer/metal interface. Thermal cyclic stresses can also be generated from the fluctuation of ambient temperatures. For example, the change in environmental temperatures is obvious when an aircraft travels across different continental regions or varies altitudes. For electronic devices, the temperature variation associated with the power on/off can reach as high as several tens of degrees. Therefore, the stress state in a hybrid composite material is not only dependent on service conditions, but also affected by the materials processing parameters. The overall stress distribution influences the fatigue crack growth behavior and the durability becomes an increasing concern. In many cases, fatigue damages in the interface region account for the majority of failures of materials.

Extensive research on the mechanical properties of metal/polymer structures has been performed [5–18]. Sekercioglu, Gulsoz and Rende [5] found that the strength of adhesively bonded cylindrical components is affected by various factors including diametrical clearance, assembly type, material properties, operating temperature, loading type, and surface roughness, among which the thickness of the interface is the most significant factor. As the bonding clearance increases, significant decreases were observed in the static and dynamic strengths. Based on the studies of carbon fiber reinforced composites bonded to steel plates, Oehlers, Liu and Seracino [6] revealed that debonding generated by shear deformation is the primary form of failure in adhesively bonded structures. Tantikom, Aizawa and Mukai [7] reported their work on symmetric to asymmetric deformation transition in regular cellular materials. The effect of adhesive bonding on such deformation mode transition was investigated under quasi-static in-plane compression loading conditions. Based on an elastic-plastic formulation through finite element (FE) analysis, a computational model was proposed for understanding the effect of various parameters on the deformation mode transition. It was found that the symmetric deformation changes to asymmetric deformation when the nominal compressive strain is increased. Cognard et al. [8] studied the behavior of thin adhesive films and tested simple composite assemblies. However, difficulty in modelling the failure of very simple joints was found due to the lack of reliable constituent input data.

Improvement of the reliability of hybrid composite materials relies on the enhancement of polymer/metal interface bonding. Various surface treatments including alkaline etching and acid pickling (applied separately or in combination with phosphoric acid anodizing) [19], plasma processing [20–23], ion beam irradiation [24–26], and coupling agent treatment [27] have been explored to examine the effect of pre-treatment on the adhesive bonding between metals and polymers. It is found that the presence of oxide and small molecules such as water in the interface region is responsible for the degradation of bonded joints [19]. Recent studies have shown that the chemical bonding at metal-polymer interface plays an important role in adhesion. Thus, the interfacial bonding and subsequent adhesion are directly influenced by the way that the interface is formed. David, Lazar and Armeanu [28] studied an aluminum/polymer joint where a thin and uniform metal sodium layer was coated on the polymer surface. The nature of the bond formation at the metal/polymer interface was investigated in view of compound formation and charge transfer between sodium and the polymer. A bonded joint was tested in terms of its strength, thermal resistance and tightness to show the interfacial properties.

Underhill and Rider [29] investigated hydrated oxide film formation on aluminum alloys immersed in warm water. Porous oxide structure was found due to the growth of hydrated oxide films on 2024 and 7075 aluminum alloys immersed in deionized water, at the temperatures of 40~50 °C for periods up to a couple of hours. In contrast with film growth studies reported for pure aluminum, the alloy systems do not appear to show an incubation period prior to hydrated oxide growth. Various characterization techniques were applied to study the properties of the oxide structure including Fourier Transform Infrared Spectroscopy (FTIR), weight gain measurements, high resolution Scanning Electron Microscopy (SEM) and Atomic Force Microscopy (AFM). It was found that the films formed at 50 °C are much thicker than those formed at 40 °C. However, the porosity of the films appears to be comparable at both temperatures. The research has suggested that a porous oxide structure is likely to be very suitable for adhesive bonding because of the increase in interface area of nanoporous structure, which results in the high shear loading capability. However, the interface nanostructure remains to be revealed by further systematic study.

POLYMER/METAL BONDED COMPOSITE STRUCTURES

In addition to making high performance hybrid composite materials, bonding composites to metals is an effective method for repairing structural defects, maintaining the load carrying capability, and extending the service life of metallic structures [30–50]. Adhesive bonding offers various advantages [43,46] and the repair of defective structures with composites has found various applications. One of the challenging aspects related to metal/polymer bonding is the long-term durability of the interface between the composites and the substrate structure. To ensure a reliable and durable bonding, materials design, stress analysis of the structures, and optimization of processing conditions have been extensively studied.

In view of materials selection, boron-epoxy, carbon-epoxy and graphite-epoxy composites have been used to form polymer composite/metal bonded structures. The thermal expansion coefficient of boron-epoxy closely matches that of most metals, which makes it one of the best candidate composite materials for structural bonding applications. Experimental work [51] showed that restoration of more than 90% strength of the original laminate is possible with the use of bonded composites. For a typical bonded configuration, a large peak in the adhesive shear strain occurs at the end of the patch. Currently, uniform stepping of the multi-layer laminate is used for bonded structures because the reduced peak adhesive shear strains lead to a smoother transition of load from metals to composites. The increase of consolidation pressure enhances the bonding of the interface.

The performance of the adhesive plays a key role in the determination of the bonding strength. Many adhesives were used to form composite/metal-bonded structures [52–56]. The adhesive bonding process requires curing at elevated temperatures, which leads to thermal residual stresses due to the thermal expansion mismatch between composite materials and metals. To decrease the thermal residual stresses in the structure, the curing cycles of high temperature adhesives were modified to temperatures lower than their standard curing conditions [57]. The optimized curing conditions were determined for each adhesive based on the bonding strength. One of the critical steps in the bonding procedure is the substrate surface preparation. To obtain sufficient shear strength for adhesive bonding, the bonding area should be thoroughly cleaned and abraded before the preparation of bonding patches. The procedures for aluminum alloy substrate preparation were given in [58,59].

Analytical work on adhesive bonding can be found in many literatures, for example [60–65]. Rose's analytical solution [61] was used by Muller *et al.* [66]. It is shown that the stress intensity factor K estimated using the Rose model provides a reasonably good connection with observed da/dN, where a is the crack length and N is the cycle of fatigue loading. Many numerical analysis techniques, such as finite element method (FEM) [67–81], boundary element method (BEM) [82–84], and finite element alternating method (FEAM) [85,86], have been proposed to the stress analysis of repaired structures. Atluri, Chow and Wang [87] applied the FEAM to model composites bonded to metals. The numerical results were compared with the experimental data.

The improved durability of bonded composite/metal structures was validated under practical application conditions. Fatigue property of notched composite/metal specimens subject to cyclic loadings was tested [88–90]. The fatigue life of cracked aluminum bonded with carbon or boron-fiber composites was extended 60 to 100 times. The fatigue life increased three times with composite patches over that of riveting metal patches to the same honeycomb panels [91]. The repair is capable of restoring residual static strength and reducing the crack growth rate by approximately two orders of magnitude [92]. Both static strength and fatigue life of the bonded plates have been significantly increased for the bonded composite patches of aircraft structures [93]. The size of adhesive-bonded region can be estimated empirically [94]. It is found that fatigue crack growth retardation is better achieved by bonding full patches to both faces of a specimen and by using a thicker composite laminate [95]. Hastie *et al.* [96] conducted photoelastic experiments on a center-cracked tension panel. The bonded repair arrested the crack for a significant number of load cycles ($N = 10^4 \sim 10^5$) and reduced the crack growth rate significantly. Jones *et al.* [97] investigated the ability of a bonded-composite doubler to restore the fatigue performance of lap joints containing multi-site damages. In all cases the repaired specimens survived more than 200,000 cycles without failure. Tay *et al.* [98] showed that if pure metallic specimens failed at about 10,000 cycles, the boron-epoxy bonded specimens survived more than 200,000 cycles. Environmental effects on the fatigue of a toughened epoxy adhesive were studied [99]. The results indicate that high temperature and high humidity tend to facilitate interface debonding and accelerate the fatigue crack growth.

The results from previous research provide useful information about materials processing and mechanical property characterization. Although metal/polymer hybrid composite materials and metal/composite repairs have been used in various fields, the relatively weak bonding between metal/ polymer interfaces still remains a problem to be solved. Studies on the interface architecturing have recently been considered. Previous work fails to consider the bonding nature and the mechanical response of the metal/polymer interface at multiscale length levels. The design and processing methodologies are limited to using the conventional techniques for fabricating fiber reinforced composite materials. Since fiber reinforced polymer/metal hybrid composite materials or repair structures contain special polymer/metal interfaces, which are highly heterogeneous in chemical, physical and mechanical properties, novel interface design methodologies are introduced to improve their performances.

NANOARCHITECTURED INTERFACES IN POLYMERIC COMPOSITES

The weakest link in hybrid composite materials or adhesive bonding structures lies in the metal/polymer interface region. As shown in Figure 1, the failure of the boron fiber/epoxy/Al hybrid composite occurs mainly by interface delamination.Figure 1(a) is the schematic of the boron fiber/epoxy bonded onto aluminum layer to form a hybrid composite structure.Figure 1(b) is a scanning electron microscopic (SEM) image showing the fracture surface morphology of the hybrid composite. From this SEM image, the "composite/Al" interface is identified as the initiation site for the failure of the hybrid composite/ structure. Conventionally available adhesives are susceptible to shear loadings because of their brittleness and weak physical bonding to metals, leading to low interface shear strength. Therefore, in order to improve the performances of metal/polymer hybrid composite materials, or bonded structural repairs, the interface bonding must be enhanced. New interface design strategies are considered to meet with the requirements. For example, nano-architecturing the interface between metal and polymer is one option. A high performance interface prototype is considered for the aluminum/epoxy system and the same principle may be extended to other metal/polymer systems.

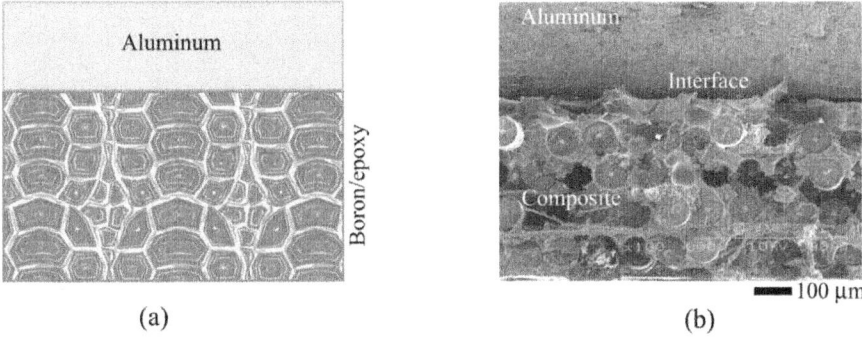

(a)   (b)

Figure 1: Composite/metal interface as the site for failure: (a) schematic of a boron/epoxy bonded to luminum substrate to form a hybrid composite structure; (b) SEM image showing the interface debonding feature.

Formation of Nanoporous Surface Structures

Self-assembled nanoporous feature in the form of anodic aluminum oxide (AAO) was prepared first on the surface of aluminum. Anodic aluminum oxide (AAO) nanoscale pores have recently been studied due to their peculiar self-organizing capability [100,101]. A two-step anodic oxidization method was used to obtain nanopores with uniform size and thin barrier layer. Aluminum alloy thin sheets were anodized on both sides using a regulated DC power supply. Before anodizing, the aluminum plates were degreased first in trichloroethylene for 2 h, followed by ultrasonic cleaning for 10 minutes in acetone. Then the samples were rinsed with methanol and distilled water, respectively. After that, the aluminum plates were etched in a 5.0 M NaOH solution at 60 °C for 5 minutes and subsequently rinsed with distilled water. Anodizing was performed in 0.3 M $H_2C_2O_4$ at ambient temperature. This first anodizing will take about 1 h. After the first anodizing, a strip-off process was carried out in a mixture solution of H_3PO_4 and H_2CrO_4. The exposed and well-ordered concave patterns on the aluminum substrate act as a self-assembled mask for the second anodizing process. The second anodizing took about 2 h. After the second anodizing, AAO templates with uniform nanopores were obtained. The depth of the nanopores was controlled in the range of 500–600 nm, and the diameter of the pores was in the range of 80–100 nm. The AAO on Al plate is schematically shown in Figure 2(a).

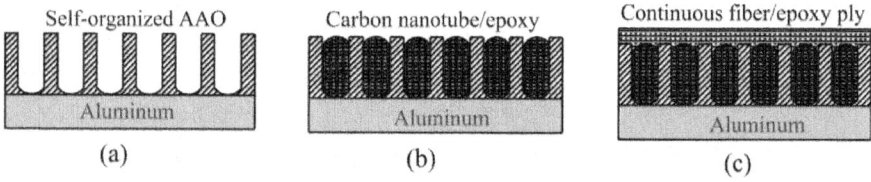

| (a) | (b) | (c) |

Figure 2: Schematic diagrams for nanoarchitecturing fiber reinforced epoxy/aluminum interface: (a) fabrication of AAO; (b) impregnation of active multi-wall carbon nanotubes/epoxy and adding nanoscaled spheres containing crack-healing resin and cure agent; (c) laminating continuous fiber reinforce epoxy plies and vacuum curing.

The nanostructured interface and hybrid composite materials are shown in Figure 2(b),(c). It is noted that the thickness of the AAO can be controlled by the oxidization time. The initially formed nanopores have the diameter about 20 nm. Pore expansion can be precisely controlled by the anodic oxidization parameters and the chemical treatment in a warm H_3PO_4 solution. The temperature, treatment time, and the concentration of the H_3PO_4 solution determine the final size of the nanopores, which may be varied between 20 nm and 200 nm. In order to control the volume fraction of the AAO nanopores, photolithography procedure may be introduced using a photo mask with microscale patterns. After the photolithography step, the surface of aluminum plate was covered by micropatterns, and oxidization to form AAO only occurs in the region without photoresist. Therefore, selective growth of AAO is achieved and the volume fraction of the AAO can be precisely controlled. The nanoscale pore formation on the surface of other metallic alloys including stainless steel and titanium alloys is also possible. These materials can be electropolished in a solution containing both HF and HNO_3 before manufacturing nanopores. By using different electrolytes and changing the electrical parameters in the anodic oxidization processes, passive films consisting NiO and Cr_2O_3 can be generated on the surface of stainless steels, while TiO_2 barrier film can be obtained on the surface of Ti-based alloys.

Addition of Active Carbon Nanotubes

Active carbon nanotubes with functional groups [102] were added into an epoxy matrix, and the modified epoxy resin containing active carbon nanotubes was introduced into the nanopores of the AAO. The active functional groups help to form strong chemical bonding both between carbon nanotubes (CNTs) and epoxy, and between epoxy and AAO. Moreover, the interface bonding is enhanced by the large specific area of the AAO, resulting in a significant improvement on the interface strength that is typically unattainable in conventional adhesive/bonding techniques.

Multi-walled and single-walled carbon nanotubes are used as additives in polymer materials to enhance the mechanical performance of the polymeric composite materials [103–116] because carbon nanotubes possess special properties [117]. Carbon nanotubes can be produced in relatively large quantities at reasonable costs using metal catalysts and either ethylene or carbon monoxide as the carbon source [118]. The structure of carbon nanotubes can be controlled through the catalyst and thermal conditions used in production. By appropriate surface treatment, carbon nanotubes present a unique, active surface so that the carbon nanotube/polymer covalent bonding can be established [119,120]. The commercially available multi-walled carbon nanotubes (MWCNTs) produced by Ahwanhnee Technology, Inc. for nano-architecturing the interface of hybrid composites were used. These tubes contain about 10 to 70 graphene layers. The diameter of the MWCNTs is in the range of 10~50 nm and the length of the tubes is in the range of 1–10 microns. Surface treatment was performed in nitric acid so that the surface of the tubes are rich in functional group of $-COOH$ as shown in Figure 3(a). The next step includes the reaction with thionyl chloride to convert the surface – $COOH$ group to acid chloride functional groups as shown in Figure 3(b). The carbon nanotubes containing acid chloride functionalities are very active to the amine cure agent for epoxy. The active carbon nanotubes were mixed with epoxy and the curing agent, as shown in Figure 3(c), secondary bonding type in the form of hydrogen bond between the AAO and the cross-linked epoxy and amine can be established. Therefore, the active carbon nanotubes are helpful to improve the interface bonding between the carbon nanotube/epoxy and epoxy/AAO. As a result, the metal/polymer interface bonding is improved.

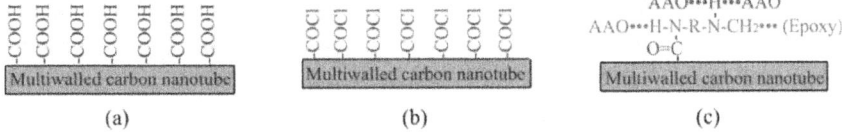

(a)   (b)   (c)

Figure 3: Illustration of preparation of active carbon nanotubes: (a) nitric acid refluxing, (b) converting $-COOH$ groups into acid chloride functional groups, (c) bonding enhancement between carbon nanotubes and AAO; carbon nanotubes and amine groups in epoxy backbones.

Addition of Nanoscale Self-Healing Capsules

Structural composites are susceptible to damage in the form of microcracks, which are usually initiated in the reinforcement/matrix interface area or deeply within the structure where detection is difficult and repair is almost impossible. Regardless of the application, microcracks are the precursors to structural failure and the ability to heal them will enable structures with extended lifetime and less maintenance. The biologically inspired, self-healing composites possess great potential for solving some of the most challenging problems of composite structural materials. Many previously reported self-repair techniques require some manual intervention [121]. Automatically healing of cracks may be accomplished by incorporating a microencapsulated healing agent and a catalytic chemical trigger in the matrix [122]: An approaching crack ruptures embedded micro- or nanocapsules, releasing a healing agent into the crack plane through capillary action. Solidification of the healing agent is then triggered by contacting with an embedded catalyst. Therefore, bonding and healing the crack faces happen.

Nanosize capsules containing self-curing resin may be added into the epoxy matrix so that active healing of the damage/small cracks in the composite materials can be achieved. Under interface debonding and crack propagation conditions, the nanoscale spheres in front of the crack tip will be broken and the curable resin such as methacrylate (MMA), and the curing agent, amine, are released as shown in Figure 4. Self-repairing effect may be achieved upon curing of the agent and the crack surfaces are bonded. Thus under the above described design, the interface structure becomes an active nanosystem, which has the function of self-healing once damage occurs. The advantage of designing such a functional active nanosystem is that the interface bonding can be substantially enhanced. Continuous fiber plies can then be stacked on the top of the multi-layered structure. Upon curing under vacuum conditions, a fiber reinforced epoxy/aluminum hybrid structure with nano-architectured interfaces is obtained. Due to the enhanced interface bonding, the novel hybrid composite structure is expected to receive wide applications, such as load carrying components with both high tensile strength, shear strength, stiffness and excellent fatigue property. The novel interface nano-architecturing strategy is also suitable for repairing defected metallic structures and can be extended to other metal/polymer systems.

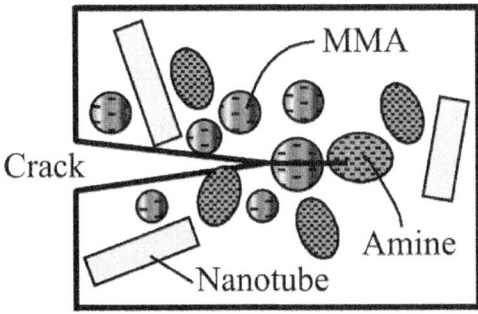

Figure 4: Illustration of active healing a crack by releasing self-curing polymers.

Self-healing polymer research has become a very active field. For example, the rheological and kinetic behaviors of self-healing agents have been studied to evaluate how fast the healing agents polymerize [123]. Urban [124] outlined the stimuli-responsive polymer networks that possess multiple functions in signaling, reorganizing and self-healing. Kavitha and Singha [125] prepared thermally amendable and self-healing homo- and copolymers by atom-transfer polymerization. Crack-healing in carbon-based materials at high temperatures has been studied [126]. Iyer and Lyon [127] reported their work on self-healing of colloidal crystals. In fiber reinforced polymeric composite materials, matrix failure is a problem. Recently, the healing behavior of a matrix crack on a carbon fiber/mendomer composite has been studied [128]. It is possible to have self-healing functions in different forms of materials or structures, for example, in films [129], coatings [130] and high elasticity block copolymers [131]. Pegoretti [132] reviewed the methods for preparing self-healing polymers and composites. Self-healing shows the increasing importance and has found various applications, especially in military [133]. Other applications such as for recycling of polymers [134] have also been studied.

NONLINEAR DAMAGE MODEL FOR POLYMERIC NA-NOCOMPOSITES

Since polymeric composites display progressive damage under external loads, nonlinearity is associated with the deformation of such materials. Nanofiber reinforced epoxies can change the wettability between the matrix and several types of reinforcements such as high strength polyethylene fibers and carbon fibers [135]. It is also possible to tailor the values of thermal expansion coefficient by controlling the content of the nanofiber in the epoxies. Therefore, nanoreinforced polymers could alleviate the residual stress problems in composite materials. For nanoreinforced composites, because of the increased toughness, nonelastic behaviors will dominate the deformation, fracture and failure processes [136,137]. Thus the prevailing linear damage models have to be modified. Nonlinear damage should be considered in nanoreinforced polymeric composites.

The nonlinear behavior of epoxies and their nanocomposites is considered as the combination of various damage events such as interfacial debonding, fiber breakage, micro/nanoscale crack propagation within the materials, and micro-yielding of the matrices. It is assumed that some of these damage events occur at low stress levels and the accumulation of damage can be revealed by the non-linear regions on stress~strain curves of the nanocomposites.

Modeling the Nonlinearity

Pervin, Zhou, Rangari *et al.* [138] proposed a nonlinear damage model to to describe the stress~strain relationship of nanocomposites. The nanocomposite materials containing the SC-15 epoxy resin and carbon nanofibers (CNFs). The contents of the nanofiber are 1 wt%, 2 wt% and 4 wt%. The characteristic parameters in the model include Young's modulus of the materials, Weibull scale parameter and Weibull shape parameter. The nonlinear relationship between the stress and strain for the epoxy nanocomposites is expressed as

$$\sigma = \epsilon E(1 - \xi) \tag{1}$$

where σ is the stress, E is the Young's modulus, ϵ is the strain, and ξ is a damage factor. ξ is considered as a Weibull function

$$\xi = P(\epsilon, E) = 1 - \exp\left[-\left(\frac{\epsilon E}{\sigma_o}\right)^m\right] \tag{2}$$

where P is the cumulative probability of failure, σ_o is the Weibull scale parameter, and m is the Weibull shape parameter. The Weibull scale parameter σ_o is the measure of the nominal strength of the composite material. With the increase in the value of σ_o the average strength of the composite increases. The shape parameter m is the measure of the scattering in strength data which reveals the dispersion of the distribution in micro/nanoscale crack length. The higher the value of m is, the less scattering of the composite strength; in other words, a larger m means more uniform and narrow flaw distribution.

Substituting Equation (2) into Equation (1) yields

$$\sigma = \epsilon E \exp\left[-\left(\frac{\epsilon E}{\sigma_o}\right)^m\right] \tag{3}$$

Rearranging Equation (3) gives

$$\frac{\sigma}{\epsilon E} = \exp\left[-\left(\frac{\epsilon E}{\sigma_\circ}\right)^m\right]$$

(4)

Taking double logarithms on both sides of Equation (4) produces

$$\ln\left[\ln\left(\frac{\sigma}{\epsilon E}\right)\right] = m\ln(\epsilon E) - m\ln(\sigma_\circ)$$

(5)

If the nonlinear behavior of the nanocomposites satisfies the Weibull distribution, Equation (5) represents a straight line. From the slope and the intercept of the line, the values for the two parameters, m and σ_0 can be determined.

Application of the Nonlinear Damage Model

Jana, Zhong and Gan [139] applied the above nonlinear model to characterize the flexural deformation behavior of reactive graphite nanofibers (r-GNFs) reinforced epoxy nanocomposite materials. It is reported that the predicted stress $vs.$ strain relations for the nanocomposites with different r-GNFs contents by the non-linear damage model are in agreement with the testing results. The m and σ_0 values for different composites calculated from experimental data using the nonlinear model of Equation (5) are listed in Table 1. From three point bending tests, it is found that the r-GNF/E-2 nanocomposite containing the epoxy and 0.3 wt% reactive graphite nanofibers (r-GNFs) has the average flexural strength around 166 MPa, while the pure epoxy specimens have an average flexural strength only around 132 MPa. For calculating the flexural strength, the following expression, Equation (6), from the beam theory was used

Table 1. Properties of nanofiber reinforced epoxy composites (data source: [139]).

Nanocomposite material type	Epoxy	r-GNF/E-1	r-GNF/E-2	r-GNF/E-3
The r-GNF content (wt%)	0	0.20	0.30	0.50
Young's modulus, E (MPa)	2770	3004	3337	3269
Flexural strength, σ_u (MPa)	132	152	166	160
Weibull shape parameter, m	26.4	8.3	3.4	4.1
Weibull scale parameter, $\sigma_\circ$ (MPa)	150	211	299	278

$$\sigma_u = \frac{3FL}{2wt^2}$$

(6)

where F is the peak load in the three point bending tests, L is the span of the specimens, w is the width of the specimens and t is the thickness of the beam specimens. The strain at failure was given by Equation (7)

$$\epsilon_u = \frac{6Dt}{L^2}$$

(7)

where D is the maximum deflection in the three point bending tests.

Referring to the data in Table 1 for the matrix and the r-GNF reinforced nanocomposites, the m value for the pure epoxy obtained from theoretical analysis using the nonlinear damage model is the highest, thus the scattering in strength for the epoxy is the lowest. The theoretical flexural stress~strain curve of the pure-epoxy is more linear compared with the experimental result, which implies that the mechanical deformation of the pure-epoxy should not be nonlinear. This is true because the epoxy is less ductile or more brittle than all the other three materials containing the reactive graphitic nanofibers. In contrast, the r-GNF/E-2 material exhibited apparent nonlinearity in the flexural test. The behavior fits well to the non-linear damage model. This can also be confirmed by the fracture surface morphology analysis, which reveals that the r-GNF/E-2 nanocomposite is more ductile than any other material listed in Table 1. The nanofiber reinforced epoxies with enhanced ductility may be used as the matrix for manufacturing high performance hybrid composites or structures. It may also be used as adhesive to make bonded joints with nanostructured interface. As discussed in Section 3., bonding composites to metallic structures is a highly cost-effective method for extending the service life of the structures. Residual stresses due to the mismatch in the coefficients of thermal expansion between metals and composites often cause premature failure of bonded structures [140]. The nanofiber reinforced epoxies as new adhesives may help to alleviate the residual stress problem because they are more ductile than the conventionally used pure epoxy adhesives.

CONCLUSIONS

Reinforcement/matrix interface plays the key role in determining the performance of advanced composite materials. To enhance the interfacial bonding, nanostructures are introduced into composite materials. Formation of nanopores on metal surface can increase the bonding strength of the metal/polymer interface. Surface treated carbon nanotubes are used in preparing nanoreinforced matrices. The nanofiber reinforced epoxies containing reactive graphitized carbon nanotubes as new adhesives can help to alleviate the residual stress problem because they are more ductile than the conventionally used pure epoxy adhesives. Finally, the progressive damage of interfaces in composites can be evaluated by nonlinear models due to the complexity of the deformation and failure processes.

ACKNOWLEDGMENTS

The author acknowledges the support of the research start-up fund from The University of Toledo.

REFERENCES

1. Afaghi-Khatibi, A; Lawcock, G; Ye, L; Mai, Y. On the fracture mechanical behaviour of fibre reinforced metal laminates (FRMLs). *Comput. Methods Appl. Mech. Engrg* 2000, *185*, 173–190.

2. Castrodeza, EM; Bastian, FL; Ipina, JEP. Critical fracture toughness, J_C and δ_C, of unidirectional fibre-metal laminates.*Thin-Walled Struct* 2003, *41*, 1089–1101.

3. Vaidya, UK; Pillay, S; Bartus, S; Ulven, CA; Grow, DT; Mathew, B. Impact and post-impact vibration response of protective metal foam composite sandwich plates. *Mater. Sci. Eng. A* 2006, *428*, 59–66.

4. Callister, WD. *Materials Science and Engineering An Introduction*, 6th ed; John Wiley & Sons, Inc: New York, NY, USA, 2005; pp. 659–662.

5. Sekercioglu, T; Gulsoz, A; Rende, H. The effects of bonding clearance and interference fit on the strength of adhesively bonded cylindrical components. *Mater. Design* 2005, *26*, 377–381.

6. Ghosh, A; Schiraldi, DA. Improving interfacial adhesion between thermoplastic polyurethane and copper foil using amino carboxylic acids. *J. Appl. Polym. Sci* 2009, *112*, 1738–1744.

7. Friedrich, J; Mix, R; Wettmarshausen, S. A new concept for adhesion promotion in metal-polymer systems by introduction of covalently bonded spacers at the interface. *J. Adhesion Sci. Technol* 2008, *22*, 1123–1143.

8. Kulkarni, RR; Chawla, KK; Vaidya, UK; Sands, JM. Thermal stresses in aluminum 6061 and nylon 66 long fiber thermoplastic (LFT) composite joint in a tailcone. *J. Mater. Sci* 2007, *42*, 7389–7396.

9. van Tijum, R; Vellinga, WP; De Hosson, JTM. Surface roughening of metal-polymer systems during plastic deformation. *Acta Mater* 2007, *55*, 2757–2764.

10. Castello, X; Estefen, SF. Limit strength and reeling effects of sandwich pipes with bonded layers. *Int. J. Mech. Sci* 2007,*49*, 577–588.

11. van Tijum, R; Vellinga, WP; De Hosson, JTM. Adhesion along metal-polymer interfaces during plastic deformation. *J. Mater. Sci* 2007, *42*, 3529–3536.

12. Li, T; Suo, Z. Ductility of thin metal films on polymer substrates modulated by interfacial adhesion. *Int. J. Solids Struct* 2007, *44*, 1696–1705.

13. Lu, N; Wang, X; Suo, ZG; Vlassak, J. Failure by simultaneous grain growth, strain localization, and interface debonding in metal films on polymer substrates. *J. Mater. Res* 2009, *24*, 379–385.

14. Zeng, F; Lv, F; Zong, RL; Wen, SP; Zhu, XY; Pan, F. Tensile properties of Cr inserted amorphous $Co_{85}Zr_9Nb_6$ films deposited on polymer substrate. *J. Alloys Compounds* 2009, *477*, 239–242.

15. Li, Y; Wang, XS; Meng, XK. Buckling behavior of metal film/substrate structure under pure bending. *Appl. Phys. Lett* 2008, *92*, 131902.

16. Oehlers, DJ; Liu, IST; Seracino, R. Shear deformation debonding of adhesively bonded plates. *Proc. Inst. Civ. Eng. Struc. Build* 2005, *158*, 77–84.

17. Tantikom, K; Aizawa, T; Mukai, T. Symmetric and asymmetric deformation transition in the regularly cell-structured materials. Part II: Theoretical study. *Int. J. Solids Struct* 2005, *42*, 2211–2224.

18. Cognard, JY; Davies, P; Gineste, B; Sohier, L. Development of an improved adhesive test method for composite assembly design. *Comp. Sci. Technol* 2005, *65*, 359–368.

19. Pires, I; Quintino, L; Miranda, RM. Performance of 2024-T3 aluminium adhesive bonded joints. *Mater. Manufact. Proc* 2005, *20*, 175–185.

20. DiFelice, RA; Dillard, JG; Yang, D. An investigation of plasma processes in titanium(iv) isobutoxide: the formation of films on Ti and Si. *Int. J. Adhesion Adhesives* 2005, *25*, 277–287.

21. Collaud, M; Groening, P; Nowak, S; Schlapbach, L. Plasma teatment of polymers-the effect of the plasma parameters on the chemical, physical and morphological states of polymer surface and on the metal-polymer interface. *J. Adhesion Sci. Technol* 1994, *8*, 1115–1127.

22. Laniog, BN; Ramos, HJ; Wada, M; Mena, MG; Flauta, RE. Surface modification of epoxy resin based PCB substrates using argon and oxygen plasmas. Proc. 2006 Int. Conf. Electronic Mater. Packaging, Shanghai, China, August 26–29, 2006; 1–3, pp. 857–862.

23. Mackova, A; Malinsky, P; Bocan, J; Svorcik, V; Pavlik, J; Stryhal, Z; Sajdl, P. Study of Ag and PE interface after plasma treatment. *Physica Status Solidi C Curr. Topics Solid State Phys* 2008, *5*, 964–967.

24. Kim, MH; Lee, KW. The effects of ion beam treatment on the interfacial adhesion of Cu/polyimide system. *Met. Mater. Int* 2006, *12*, 425–433.

25. Zaporojtchenko, V; Zekonyte, J; Faupel, F. Effects of ion beam treatment on atomic and macroscopic adhesion of copper to different polymer materials. *Nucl. Instrum. Meth. Phys. Res. B* 2007, *265*, 139–145.

26. Ratchev, BA; Was, GS; Booske, JH. Ion beam modification of metal-polymer interfaces for improved adhesion. *Nucl. Instrum. Meth. Phys. Res. B* 1995, *106*, 68–73.

27. Ghosh, A; Schiraldi, DA. Improving interfacial adhesion between thermoplastic polyurethane and copper foil using amino carboxylic acids. *J. Appl. Polym. Sci* 2009, *112*, 1738–1744.

28. David, E; Lazar, A; Armeanu, A. Surface modification of polytetrafluoroethylene for adhesive bonding. *J. Mater. Proc. Technol* 2004, *157*, 284–289.

29. Underhill, PR; Rider, AN. Hydrated oxide film growth on aluminium alloys immersed in warm water. *Surf. Coat. Technol* 2005, *192*, 199–207.

30. Shin, KC; Lee, JJ. Effects of thermal residual stresses on failure of co-cured lap joints with steel and carbon fiber-epoxy composite adherends under static and fatigue tensile loads. *Composites Part A: Appl. Sci. Manufact* 2006, *37*, 476–487.

31. Cognard, J. Some recent progress in adhesion technology. *Comptes Rendus Chimie* 2006, *9*, 13–24.

32. Barroso, A; Paris, F; Mantic, V. Representativity of the singular stress state in the failure of adhesively bonded joints between metals and composites. *Comp. Sci. Technol* 2009, *69*, 1746–1755.

33. Feraboli, P; Masini, A. Development of carbon/epoxy structural components for a high performance vehicle.*Composites Part B: Eng* 2004, *35*, 323–330.

34. Woerdeman, DL; Emerson, JA; Giunta, RK. JKR contact mechanics for evaluating surface contamination on inorganic substrates. *Int. J. Adhesion Adhesives* 2002, *22*, 257–264.

35. Roche, AA; Bouchet, J; Bentadjine, S. Formation of epoxy-diamine/metal interphases. *Int. J. Adhesion Adhesives* 2002,*22*, 431–441.

36. Ageorges, C; Ye, L; Hou, M. Advances in fusion bonding techniques for joining thermoplastic matrix composites: A review. *Composites Part A: Appl. Sci. Manufact* 2001, *32*, 839–857.

37. Rao, VV; Singh, R; Malhotra, SK. Residual strength and fatigue life assessment of composite patch repaired specimens. *Composites Part B* 1999, *30*, 621–627.

38. Muller, R; Fredell, R. Analysis of multiple bonded patch interaction simple design guidelines for multiple bonded repairs in close proximity. *Appl. Comp. Mater* 1999, *6*, 217–237.

39. Klug, JC; Sun, CT. Large deflection effects of cracked aluminum plates repaired with bonded composite patches.*Comp. Struct* 1998, *42*, 291–296.

40. Fredell, R; Guijt, C; Mazza, J. An integrated bonded repair system: A reliable means of giving new life to aging airframes. *Appl. Comp. Mater* 1999, *6*, 269–277.

41. Kumar, AM; Hakeem, SA. Optimum design of symmetric composite patch repair to centre cracked metallic sheet.*Comp. Struct* 2000, *49*, 285–292.

42. Wang, WC; Hsu, JS. Investigation of the size effect of composite patching repaired on edge-cracked plates. *Comp. Struct* 2000, *49*, 415–423.

43. Christian, TF, Jr; Hammond, DO; Cochran, JB. Composite material repairs to metallic airframe components. *J. Aircraft* 1992, *29*, 470–476.

44. Umamaheswar, TV; Singh, R. Modelling of a patch repair to a thin cracked sheet. *Eng. Fract. Mech* 1999, *62*, 267–289.

45. Denney, JJ; Mall, S. Characterization of disbond effects on fatigue crack growth behavior in aluminum plate with bonded composite patch. *Eng. Frac. Mech* 1997, *57*, 507–509.

46. Baker, AA; Rose, RLF; Walker, KF. Repair substantiation for a bonded composite repair to F111 lower wing skin. *Appl. Comp. Mater* 1999, *6*, 251–267.

47. Jones, R; Smith, WR. Continued airworthiness of composite repairs to primary structures for military aircraft. *Comp. Struct* 1995, *33*, 17–26.

48. Fernandes, PJL; Jones, DRH. The effects of loading variables on fatigue-crack growth in liquid-metal environments.*Int. J. Fatigue* 1995, *17*, 501–505.

49. Siddaramaiah, JVA; Gowri, ANS; Leena, B; Hemmige, VV; Deepa, S. Tensile behavior of composite patched cracked metallic structures. *J. Appl. Polym. Sci* 1998, *68*, 2063–2068.

50. Wang, CH; Rose, LRF; Baker, AA. Modelling of the fatigue growth behaviour of patched cracks. *Int. J. Fract* 1998, *88*, L65–L70.

51. Fredell, R; Vanbarneveld, W; Vlot, A. Analysis of composite crack patching of fuselage structures- high patch elastic modulus is not the whole story. Moving Forward with 50 Years of Leadership in Adv. Mater.-39th Int. SAMPE Sym. Exhibition, Anaheim, California, USA, April 11–14, 1994; 39, pp. 610–623.

52. Chow, WT; Atluri, SN. Composite patch repairs of metal structures: Adhesive nonlinearity, thermal cycling, and debonding. *AIAA J* 1997, *35*, 1528–1535.

53. Naboulsi, S; Mall, S. Fatigue crack growth analysis of adhesively repaired panel using perfectly and imperfectly composite patches. *Theor. Appl. Fract. Mech* 1997, *28*, 13–28.

54. Schubbe, JJ; Mall, S. Investigation of a cracked thick aluminum panel repaired with a bonded composite patch. *Eng. Fract. Mech* 1999, *63*, 305–323.

55. Ong, CL; Shen, SB. Some results on metal and composite patch Reinforcement of aluminum honeycomb pannel. *Theor. Appl. Fract. Mech* 1991, *16*, 145–153.

56. Chalkley, P; Baker, A. Development of a generic repair joint for certification of bonded composite repairs. *Int. J Adhesives* 1999, *19*, 121–132.

57. Dong, CS. Modeling the process-induced dimensional variations of general curved composite components and assemblies. *Composites Part A: Appl. Sci. Manuf* 2009, *40*, 1210–1216.

58. Aglan, H; Gan, YX; Wang, QY; Kehoe, M. Design guidelines for composite patches bonded to cracked aluminum substrates. *J. Adhesion Sci. Technol* 2002, *16*, 197–211.

59. Salehi-Khojin, A; Zhamu, A; Zhong, W; Gan, Y. Effects of patch layer and loading frequency on fatigue fracture behavior of aluminum plate repaired with a boron/epoxy composite patch. *J. Adhesion Sci. Technol* 2006, *20*, 107–123.

60. Nassar, SA; Virupaksha, VL. Effect of adhesive thickness and properties on the biaxial interfacial shear stresses in bonded joints using a continuum mixture model. *J. Eng. Mater. Technol* 2009, *131*, 021015.

61. Rose, LRF. A cracked plate repaired by bonded reinforcements. *Int. J Fract* 1982, *18*, 135–144.

62. Rose, LRF. An application of the inclusion analogy for bonded Reinforcements. *Int. J. Solids Struct* 1981, *17*, 827–838.

63. Lam, YC; Zhu, CS. Analytical techniques for bonded repairs. *Fract Stregth Solids Part 1–2* 1998, *145–149*, 543–552.

64. Xiong, Y. An analytical model for stress analysis and failure prediction of bonded joints with tapered edges. *Trans. Can. Soc. Mech. Eng* 1998, *22*, 143–155.

65. Renton, WJ; Vinson, JR. Analysis of adhesively bonded joints between panels of composite-materials. *J. Appl. Mech*1977, *44*, 101–106.

66. Muller, R; Fredell, R; Guijt, C. Experimental verification of Rose's constant K solution in bonded crack patching. *Appl. Comp. Mater* 1999, *6*, 205–216.

67. Lena, MR; Klug, JC; Sun, CT. Composite patches as reinforcements and crack arrestors in aircraft structures. *J. Aircraft*1998, *35*, 318–323.

68. Sun, CT; Klug, J; Arendt, C. Analysis of cracked aluminum plates repaired with bonded composite patches. *AIAA J*1996, *34*, 369–374.

69. Monaco, A; Sinke, J; Benedictus, R. Experimental and numerical analysis of a beam made of adhesively bonded tailor-made blanks. *Int. J. Adv. Manuf. Technol* 2009, *44*, 766–780.

70. Kim, WS; Lee, JJ. Fracture characterization of interfacial cracks with frictional contact of the crack surfaces to predict failures in adhesive-bonded joints. *Eng. Fract. Mech* 2009, *76*, 1785–1799.

71. Lin, CC; Chu, RC; Lin, YS. A finite-element model for single-sided crack patching. *Eng. Fract. Mech* 1993, *46*, 1005–1021.

72. Kam, TY; Chu, KH; Tsai, YC. Fatigue of cracked plates repaired with single-sided composite patches. *AIAA J* 1998, *36*, 645–650.

73. Kam, TY; Tsai, YC; Chu, KH; Wu, JH. Fatigue analysis of cracked aluminum plates repaired with bonded composite patches. *AIAA J* 1998, *36*, 115–118.

74. Naboulsi, S; Mall, S. Characterization of fatigue crack growth in aluminium panels with a bonded composite patch.*Comp. Struct* 1997, *37*, 321–334.

75. Naboulsi, S; Mall, S. Methodology to analyse aerospace structures repaired with a bonded composite patch. *J. Strain Analy. Eng. Design* 1999, *34*, 395–412.

76. Naboulsi, S; Mall, S. Nonlinear analysis of bonded composite patch repair of cracked aluminum panels. *Comp. Struct*1998, *41*, 303–313.

77. Ratwani, MM. Analyis of cracked, adhesively bonded laminated structures. *AIAA J* 1979, *17*, 988–994.

78. Chau, RWT; Lee, SWR. Computational modeling and analysis of a center-cracked panel repaired by bonded composite patch. *Fract. Strength Solids, Part 1–2* 1998, *145–149*, 601–606.

79. Paul, J; Bartholomeusz, RA; Jones, R; Ekstrom, M. Bonded composite repair of cracked load-bearing holes. *Eng. Fract. Mech* 1994, *48*, 455–461.

80. Chue, C; Liu, TJ. The effects of laminated composite patch with different stacking sequences on bonded repair. *Comp. Eng* 1995, *5*, 223–230.

81. Callinan, RJ; Galea, SC; Sanderson, S. Finite element analysis of bonded repairs to edge cracks in panels subjected to acoustic excitation. *Comp. Struct* 1997, *38*, 649–660.

82. Chue, CH; Chou, WC; Liu, TJC. The effects of size and stacking sequence of composite laminated patch on bonded repair for cracked hole. *Appl. Comp. Mater* 1996, *3*, 355–367.

83. Tarn, JQ; Shek, KL. Analusis of cracked plates with a bonded patch. *Eng. Fract. Mech* 1991, *40*, 1055–1065.

84. Young, A; Rooke, DP; Cartwright, DJ. Analysis of patched and stiffened cracked panels using the boundary element method. *Int. J. Solids Struct* 1992, *29*, 2201–2216.

85. Salgado, NK; Aliabadi, MH. The boundary element analysis of cracked stiffened sheets, reinforced by adhesively bonded patches. *Int. J. Numer. Method Eng* 1998, *42*, 195–217.

86. Zadpoor, AA; Sinke, J; Benedictus, R. The mechanical behavior of adhesively bonded tailor-made blanks. *Int. J. Adhesion Adhesives* 2009, *29*, 558–571.

87. Atluri, SN; Chow, WT; Wang, L. Computational mechanics of interfacial fracture for composite debonding; And of elastoplastic fracture with multi-site-damage. *Adv Fract Res* 1997, *1–6*, 1885–1897.

88. Nahas, MN. Experimental investigation of fatigue of cracked aluminum specimens repaired with fiber composite patches. *J. Reinf. Plast. Comp* 1992, *11*, 932–938.

89. Ong, CL; Chu, RC; Ko, TC; Shen, SB. Composite patch reinforcement of cracked aircraft upper longeron-analysis and specimen simulation. *Theor. Appl. Fract. Mech* 1990, *14*, 13–26.

90. Ong, CL; Shen, SB. The reinforced effect of composite patch repairs on metalic aircraft structures. *Int. J. Adhesion and Adhesives* 1992, *12*, 19–26.

91. Sabelkin, V; Mall, S; Hansen, MA; Vandawaker, RM; Derris, M. Investigation into cracked aluminum plate repaired with bonded composite patch. *Comp. Struct* 2007, *79*, 55–66.

92. Chester, RJ; Walker, KF; Chalkley, PD. Adhesively bonded repairs to primary aircraft structure. *Int. J. Adhesion Adhesives* 1999, *19*, 1–8.

93. Clark, RJ; Romilly, DP. Bending of bonded composite repairs for aluminum aircraft structures: A design study. *J. Aircraft* 2007, *44*, 2012–2025.

94. Rao, VV; Singh, R; Malhotra, SK. Residual strength and fatigue life assessment of composite patch repaired specimens. *Composites Part B: Eng* 1999, *30*, 621–627.

95. Alawi, H; Saleh, IE. Fatigue crack-growth retardation by bonding patches. *Eng. Fract. Mech* 1992, *42*, 861–868.

96. Hastie, RL; Fredell, R; Dally, JW. A photoelastic study of crack repair. *Exp. Mech* 1998, *38*, 29–36.

97. Jones, R; Bartholomeusz, R; Kayer, R; Roberts, J. Bonded-composite repair of representative multisite damage in a full-scale fatigue-test. *Theor. Appl. Fract Mech* 1994, *21*, 41–49.

98. Tay, TE; Chau, FS; Er, CJ. Bonded boron-epoxy composite repair and reinforcement of cracked aluminium structures.*Comp. Struct* 1996, *34*, 339–347.

99. Butkus, J. Considering environmental conditions in the design of bonded structures: A fracture mechanics approach.*Fatig. Fract. Eng. Mater. Sruct* 1998, *21*, 465–478.

100. Yang, Z; Huang, Y; Dong, B; Li, H; Shi, S. Densely packed single-crystal $Bi_2Fe_4O_9$ nanowires fabricated from a template-induced solgel route. *J. Solid State Chem* 2006, *179*, 3324–3329.

101. Zhao, Y; Chen, M; Zhang, Y; Xu, Z; Liu, W. A facile approach to formation of through-hole porous anodic aluminum oxide film. *Mater. Lett* 2005, *59*, 40–43.

102. Zhong, WH; Li, J; Xu, LR; Lukehart, CM. Graphitic carbon nanofiber (GCNF)/polymer materials, II. GCNF/epoxy monoliths using active oxydianiline linker molecules and the effect of nanofiber reinforcement on curing conditions.*Polym. Compos* 2005, *26*, 128–135.

103. Corral, EL; Cesarano, J; Shyam, A; Lara-Curzio, E; Bell, N; Stuecker, J; Perry, N; Di Prima, M; Munir, Z; Garay, J; Barrera, EV. Engineered nanostructures for multifunctional single-Walled carbon nanotube reinforced silicon nitride nanocomposites. *J. Am. Ceram. Soc* 2008, *91*, 3129–3137.

104. Chipara, M; Lozano, K; Wilkins, R; Barrera, EV; Pulikkathara, MX; Penia-Para, L. ESR investigations on polyethylene-single wall carbon nanotube composites. *J. Mater. Sci* 2008, *43*, 1228–1233.

105. Zhu, J; Imam, A; Crane, R; Lozano, K; Khabashesku, VN; Barrera, EV. Processing a glass fiber reinforced vinyl ester composite with nanotube enhancement of interlaminar shear strength. *Comp. Sci. Technol* 2007, *67*, 1509–1517.

106. McIntosh, D; Khabashesku, VN; Barrera, EV. Benzoyl peroxide initiated in situ functionalization, processing, and mechanical properties of single-walled carbon nanotube-polypropylene composite fibers. *J. Phys. Chem. C* 2007, *111*, 1592–1600.

107. McIntosh, D; Khabashesku, VN; Barrera, EV. Nanocomposite fiber systems processed from fluorinated single-walled carbon nanotubes and a polypropylene matrix. *Chem. Mater* 2006, *18*, 4561–4569.

108. Shofner, ML; Khabashesku, VN; Barrera, EV. Processing and mechanical properties of fluorinated single-wall carbon nanotube-polyethylene composites. *Chem. Mater* 2006, *18*, 906–913.

109. Wilkins, R; Pulikkathara, MX; Khabashesku, VN; Barrera, EV; Vaidyanathan, RK; Thibeault, SA. Ground-based space radiation effects studies on single-walled carbon nanotube materials. *Mater. Space Appl* 2005, *851*, 267–278.

110. Zeng, Q; Bayazitoglu, Y; Zhu, J; Wilson, K; Imam, MA; Barrera, EV. Coating of SWNTs with nickel by electroless plating method. The 5th Pacific Rim Int. Conf. Adv. Mater. Processing Proceedings, Beijing, China, November 2–5, 2005; 475–479, pp. 1013–1018.

111. Shofner, ML; Peng, HQ; Gu, ZN; Khabashesku, VN; Margrave, JL; Barrera, EV. Mechanical properties of polyethylene containing defunctionalized single wall carbon nanotubes. *Mech. Prop. Nanostruct. Mater. Nanocomp* 2004, *791*, 373–378.

112. Zhu, J; Peng, HQ; Rodriguez-Macias, F; Margrave, JL; Khabashesku, VN; Imam, AM; Lozano, K; Barrera, EV. Reinforcing epoxy polymer composites through covalent integration of functionalized nanotubes. *Adv. Funct. Mater* 2004, *14*, 643–648.

113. Yang, SY; Castilleja, JR; Barrera, EV; Lozano, K. Thermal analysis of an acrylonitrile-butadiene-styrene/SWNT composite. *Polym. Degrad. Stability* 2004, *83*, 383–388.

114. Shofner, ML; Rodriguez-Macias, FJ; Vaidyanathan, R; Barrera, EV. Single wall nanotube and vapor grown carbon fiber reinforced polymers

processed by extrusion freeform fabrication. *Composites Part A Appl. Sci. Manuf* 2003, *34*, 1207–1217.

115. Pulikkathara, MX; Shofner, ML; Wilkins, RT; Vera, JG; Barrera, EV; Rodriguez-Macias, FJ; Vaidyanathan, RK; Green, CE; Condon, CG. Fluorinated single wall nanotube/polyethylene composites for multifunctional radiation protection.*Nanomater. Struct. Appl* 2003, *740*, 365–370.

116. Barrera, EV. Key methods for developing single-wall nanotube composites. *JOM-J. Min. Met. Mater. Soc* 2000, *52*, A38–A42.

117. Saito, R; Dresselhaus, G; Dresselhaus, MS. *Physical Properties of Carbon Nanotubes*; Imperial College Press: London, UK, 1998; pp. 37–52.

118. Meyyappan, M. *Carbon Nanotubes Science and Applications*; CRC Press: Boca Raton, 2005; pp. 21–74.

119. Liao, YH; Marietta-Tondin, O; Liang, ZY; Zhang, C; Wang, B. Investigation of the dispersion process of SWNTs/SC-15 epoxy resin nanocomposites. *Mater. Sci. Eng. A* 2004, *385*, 175–181.

120. Xie, J; Wong, P. Nanofluids containing multiwalled carbon nanotubes and their enhanced thermal conductivities. *J. Appl. Phys* 2003, *94*, 4967–4968.

121. Sillikas, N; Al-Kheraif, A; Watts, DC. Influence of P/L ratio and peroxide/amine concentrations on shrinkage-strain kinetics during setting of PMMA/MMA biomaterial formulations. *Biomaterials* 2005, *26*, 197–204.

122. White, SR; Sottos, NR; Geubelle, PH; Moore, JS; Kessler, MR; Sriram, SR; Brown, EN; Viswanathan, S. Autonomic healing of polymer composites. *Nature* 2001, *409*, 794–796.

123. Liu, X; Sheng, X; Lee, JK; Kessler, MR; Kim, JS. Rheokinetic evaluation of self-healing agents polymerized by Grubbs catalyst embedded in various thermosetting systems. *Comp. Sci. Technol* 2009, *69*, 2102–2107.

124. Urban, MW. Stratification, stimuli-responsiveness, self-healing, and signaling in polymer networks. *Progr. Polym. Sci* 2009, *34*, 679–687.

125. Kavitha, AA; Singha, NK. "Click Chemistry" in tailor-made polymethacrylates bearing reactive furfuryl functionality: A new class of self-healing polymeric material. *ACS Appl. Mater. Interf* 2009, *1*, 1427–1436.

126. Yun, J; Choi, Y; Lee, H. Crack-healing capability and high temperature oxidation resistance of multilayer coatings for carbon-carbon composites. *J. Ceram. Proc. Res* 2009, *10*, 340–343.

127. Iyer, AS; Lyon, LA. Self-healing colloidal crystals. *Angewandte Chemie-Int. Ed* 2009, *48*, 4562–4566.

128. Park, JS; Kim, HS; Hahn, HT. Healing behavior of a matrix crack on a carbon fiber/mendomer composite. *Comp. Sci. Technol* 2009, *69*, 1082–1087.

129. Ensslin, S; Moll, KP; Haefele-Racin, T; Maeder, K. Safety and robustness of coated pellets: self-healing film properties and storage stability. *Pharm. Res* 2009, *26*, 1534–1543.

130. Wouters, M; Craenmehr, E; Tempelaars, K; Fisher, H; Stroeks, N; van Zanten, J. Preparation and properties of a novel remendable coating concept. *Progr. Org. Coat* 2009, *64*, 156–162.

131. Chipara, MD; Chipara, M; Shansky, E; Zaleski, JM. Self-healing of high elasticity block copolymers. *Polym. Adv. Technol* 2009, *20*, 427–431.

132. Pegoretti, A. The way to autonomic self-healing polymers and composites. *Expr. Polym. Lett* 2009, *3*, 62–62.

133. Dubey, R; Shami, TC; Rao, KUB. Microencapsulation technology and applications. *Defence Sci. J* 2009, *59*, 82–95.

134. Zhang, Y; Broekhuis, AA; Picchioni, F. Thermally self-healing polymeric materials: the next step to recycling thermoset polymers? *Macromolecules* 2009, *42*, 1906–1912.

135. Neema, S; Salehi-Khojin, A; Zhamu, A; Zhong, WH; Jana, S; Gan, YX. Wettability of nano-epoxies to UHMWPE fibers.*J. Colloid Interf. Sci* 2006, *299*, 332–341.

136. Jana, S; Zhamu, A; Zhong, WH; Gan, YX. Experimental evaluation of adhesion property of UHMWPE fibers/nano-epoxy by a pullout test. *J. Adhesion* 2006, *82*, 1157–1175.

137. Jana, S; Zhamu, A; Zhong, WH; Gan, YX; Stone, JJ. Effect of reactive graphitic nanofibers (r-GNFs) on tensile behavior of UHMWPE fiber/nano-epoxy bundle composites. *Mater. Manuf. Proc* 2008, *23*, 102–110.

138. Pervin, F; Zhou, Y; Rangari, VK; Jeelani, S. Testing and evaluation on the thermal and mechanical properties of carbon nano fiber reinforced SC-15 epoxy. *Mater. Sci. Eng. A* 2005, *405*, 246–253.

139. Jana, S; Zhong, W; Gan, YX. Characterization of the flexural behavior of a reactive graphitic nanofibers reinforced epoxy using a non-linear damage model. *Mater Sci Eng A* 2007, *445–446*, 106–112.

140. Zhong, WH; Zhamu, A; Aglan, A; Stone, J; Gan, YX. *J. Adhesive Sci. Technol* 2005, *19*, 1113–1128.

Chapter 7

GROWTH REINFORCING COMPOSITE MATERIALS FROM LIQUIDUS PHASE: MECHANICAL AND MICROSTRUCTURAL PARAMETERS RELATIONSHIP ESSENTIALLY EVINCING THE PREDOMINANCE OF AN AKIN MASS COMPOSITION OVER THE DOMAIN OF COMPOSITIONS

B.L. Sharma and Parshotam Lal
Department of Chemistry, University of Jammu, India

INTRODUCTION

Composite materials are heterogeneous systems consisting of two or more physically distinct and mechanically separable materials belonging to different space groups, and are in equilibrium with a single liquidus phase. The essence of composite materials lies in the concept that their properties must essentially be superior, and possibly unique in some specific respects, to those of the constituent phases, particularly if these are geometrically oriented to one another in some periodic or anisotropic manner (Sharma et al., 2008, 2009). Composite materials designed by anisotropic growth process, can offer advanced possibilities as reinforced materials for the constructions and further development of supersonic aircraft, space vehicles, high pressure tanks, for which strength properties are required that cannot be provided by the existing homogeneous materials (Ashbee & Woishnis, 1993). Besides mechanical fields, these materials are equally promising in a variety of other fields (Hull & Clyne, 2006). For example, most metallic permanent magnets are made of alnico alloys, obtained by unidirectional solid-state decomposition under cooling conditions in an applied magnetic field. The rod composite structure of quasi binary NiSb-InSb (Caram et al., 1990) obtained by unidirectional solidification from its homogeneous eutectic melt, finds a representative example in electronics. Currently, the femoral stem is constructed from Co-Cr-Mo, Co-Ni-Cr-Mo and Ti-Al-V alloys (Callister, 2006; Mallick, 1993 & Pillar, 1984) in preference to stainless steels. The advantages of composite materials appear when the specific modulus (modulus per unit weight) and specific strength (strength per unit weight) are considered. The higher specific modulus and specific strength of composite materials means that the weight of constituent phases can be reduced (Callister, 2006 &Courtney, 2000). This is a factor of greater importance in moving constituent phases in all forms of transport where reductions in weight result in greater efficiency and energy savings (Callister & Rethwisch, 2008).

The essence of anisotropic growth applied to composite materials is the ability to put strong stiff fibers in the right place, in the right orientation with right volume fraction. Implicit in the present searching approach is the concept that while in designing the composite material by variable anisotropic growth, the solidification aspect is also related to discover the final product as an acceptable analog for manufacturing the engineering product. Because of the complex nature of the phenomena involved in an anisotropic growth process from the melt, a considerable judgment is required to assess both theoretical and experimental observations, since the composite materials that can be developed by this process exhibit a large diversity of micro-morphologies. However, a physical understanding of the mechanical properties of composite materials as a basis for the improvement of the properties by controlling the growth rates from the melt can readily be distinguished when examined in an optical or electronic microscope. In view of watching the growth process visually, binary organic eutectic systems being transparent are a most suitable option for investigation, since metals are opaque and the observations had to be made on external surfaces. A very interesting finding of the current investigation is an akin mass composition (the composition whereat masses of the constituent members would almost overlap each other) that predominates the domain of compositions, since anisotropic mechanical parameters obeying the Weibull probability distribution curve evince the predominance, particularly at moderate anisotropic growth rate. Likewise, of greater interest is the moderate anisotropic growth in the domain of solidification modes as the modal products structured by the growth process comprise microstructures which are three-to-four fold superior in mechanical strength to their isotropic growth performed in an ice bath (273 K) and many fold to their respective individual components irrespective of the growth pattern. Vickers microhardness offers supporting evidence to the essence of an identical form of the Weibull distribution curve demonstrating the strength-growth relationship. The liquidus structure encompassing the entire composition of binary organic eutectic systems is characterized by rheological view point comprising of viscosity and surface tension measurements at different temperatures, which manifests specific interactions occurring in the eutectic compositional melt that are complimentary supported by thermodynamic analysis in terms of excess functions. The experimental data on kinetics and anisotropic crystal growth from the melt verify nucleation theory and evince the essence of dislocation mechanism. Thermal and X-ray diffraction studies of binary composite materials are accomplished in order to ascertain thermal stability, purity, composition, enthalpies of fusion and mechanical combination of homogeneous materials.

A series of binary organic eutectic systems, viz., naphthalene-o-nitrophenol; α-naphthol-naphthalene; diphenylamine-α-naphthol; benzil-diphenyl; benzoic acid-benzamide; benzoic acid-naphthalene; naphthalene-phenanthrene; α-naphthylamine-diphenylamine and binary metallic eutectic systems, namely, Sn-Cd; Cd-Bi; Pb-Sn and Pb-Bi, were selected for the investigation. The choice of homogeneous materials in aforementioned systems is restricted to the methodical considerations: low temperatures of melting, purity and the possibility of precisely controlling and measuring temperatures.

EXPERIMENTAL DETAILS

Materials and Methodology

A great precautious care was taken to the selection of utmost pure homogeneous materials, since organic materials are less pure and even small portion of soluble impurities can essentially influence the crystal growth. Consequently, preference was given to analytical reagent, especially of The British Drug House Limited and Merck for 99.9% pure organic materials (naphthalene, o-nitrophenol, α-naphthol, diphenylamine, benzil, diphenyl, benzoic acid, benzamide, phenanthrene and α-naphthylamine) and Alfa Aesar for 99.999% pure metals (Pb, Sn, Bi and Cd). The melting temperatures and enthalpies of fusion of the experimental homogeneous materials were determined by thaw-melt technique and thermal analysis (DSC Linseis STA PT-1000) and found consistent with the literal values (Lide, 2009).

Diagrams of State

Mixtures of the constituent materials were prepared covering the whole of the mole fraction composition range (binary organic systems) and weight percent composition converted into mole fraction composition range (binary metallic systems). The homogeneity in mixtures was ascertained by heat-chill method. The solidus-liquidus equilibrium curves of binary organic systems were studied by thaw-melt method, while those of binary metallic systems by thermal analysis.

Heterogeneous Nucleation

Nucleation studies of binary organic systems being transparent in nature were performed following the procedure referred earlier (Sharma et al., 2004a). A nearly constant amount (4.0 g) of each composition of the systems was carefully taken into different pyrex tubes of uniform internal diameters. The sample tubes were sealed and successively undertaken for spontaneous nucleation observations in a paraffin bath manipulated at the temperature ~30K higher than the melting temperature of each specimen. The bath temperature attained by the tube was ensured by thermometric technique before cooling process for spontaneous crystallization. The melt-nucleation process for each sample tube was recycled for several observations (minimum six runs) necessary in comparative estimate of undercooling. Maximum care was taken to perform undercooling experiments for different compositional melts under almost identical conditions by keeping the volume of the melt and the rate of cooling nearly the same in all the cases.

Anisotropic Growth Velocity

The experimental technique for determining the anisotropic growth velocity from different compositional melts of binary organic systems was similar to that adopted earlier (Sharma, 2003). The velocity of anisotropic growth from a definite compositional melt was measured in a transverse direction along the length 10×10^{-2} m and internal diameter of 6×10^{-2} m of the pyrex tube with terminal right-angled bends. The experimental tube was placed in a paraffin thermostat kept constant to ±0.01 K for experimental observations at several degrees of undercooling. The rate of advance of the crystal boundary was recorded by stop watch. Several observations (minimum six runs) of anisotropic growth velocity were recorded for each compositional melt at the selected experimental supercooling. The results are found less accurate from the eutectic melts of each system, since within the observation time the growing crystal front moved only a very small distance which could not be ascertained very accurately.

Rheological Measurements

Viscosities of different compositional melts representing the complete mole fraction composition range of each binary organic system, were measured relative to the literal viscosity values of reference liquid p-xylene (Lide, 2009) in a paraffin bath with an accuracy of ±0.05 K in the temperature range 391-408 K. The viscometer employed for the purpose was modified as shown in Fig.A. Likewise, an attempt was made for the first time to measure surface tension of binary α-napthylamine-diphenylamine eutectic melt system related to the reference liquid p-xylene (Lide, 2009 & Edward, 1930) in the aforementioned bath in the temperature range 323-373 K using modified stalagmometer (Fig.B) compatible with the suitability of the experiment. The viscometer and stalagmometer are essentially modified by the side tube 'S' connecting the bulb 'B' right below the capillary of the viscometer to its upper mark 'X', and the bulb 'B' of the stalagmometer to the upper mark 'X' in the light of the following:

1. the excess pressure of the melt increasing the kinetic energy of the flow is made to escape through the side tube 'S' leaving the experimental volume of the melt to flow freely down the capillary under its own gravity. This minimizes kinetic energy correction.

2. the bulb 'B' is an important component of the viscometer in connection with the elimination of upward thrust due to surface tension which could resist the free flow of the melt. The bulb 'B$_1$' of the stalagmometer would also overcome in nipping the resistance.

3. the advantage of the modified stalagmometer over other stalagmometer lies in the concept that the relative surface tension measurements of low temperature solid materials in general and binary or ternary solid mixtures in particular can easily be undertaken at different temperatures in an air oven or a suitable bath to assess the specific interactions in their respective melts' surface.

Growth Technique

Anisotropic Crystal Form

Pyrex tubes of nearly the same dimensions were selected as dies to achieve anisotropic rod form of crystals from different compositional melts in the following experimental setup. An experimental pyrex tube containing half-full melt of freshly prepared eutectic or non-eutectic phase, was vertically clamped to the centre of an empty graduated beaker (volume capacity 1dm^3) arranged in an oven set at a temperature ~30 K higher than the melting temperature of the experimental specimen. The molten mass was nucleated by circulating water in case of organic melt, and silicone oil in case of metallic melt from their respective reservoirs manipulated to perforate at ~300 K. The melt in the tube started nucleating when the rising level of water or silicone oil as the case may be, just touched the bottom of the tube creating sharp temperature gradient at the solidus-liquidus interface that resulted in the occurrence of melt-growth process anisotropically towards longitudinal direction along the length of the tube. In this manner, each compositional melt in a separate pyrex tube was grown to crystal form at 18 different intervals in the experimental time range 5-45 min. The crystal sample grown anisotropically was removed from an experimental pyrex tube by warming its section in both longitudinal and transverse directions just to glide the solidus mass to its easier exit.

Isotropic Crystal Form

Nuclei of critical size almost grew out instantaneously in all directions when a pyrex tube containing the melt at ~30 K higher than its freezing temperature was dropped vertically into an ice bath maintained at ~273 K. This particular mode of solidification is presumed to be time independent and thereby zero order reference growth. The solidified crystal from the tube was removed following the aforementioned procedure. A good many crystal samples of eutectic and akin mass eutectic compositions, and the constituent materials were grown isotropically and anisotropically from their respective melts at different but nearly constant growth rates determined by circulating approximately the same volume of water or silicone oil for each experimental interval. An interesting feature of time variable growth is the yield of moderate anisotropic growth velocity (~2.90 X 10^{-7} m^3s^{-1}) determined by setting the flow-interval of reservoir liquid at ~5 X 10^{-4} m^3 for 30 min. At this moderate solidification rate, the growth from an akin mass compositional melt of each experimental binary organic system is found to be more efficient in promoting its mechanical properties which are superior, as revealed by the experimental evidences in the present work, to those of the eutectic composite and constituent materials grown under the same environmental conditions.

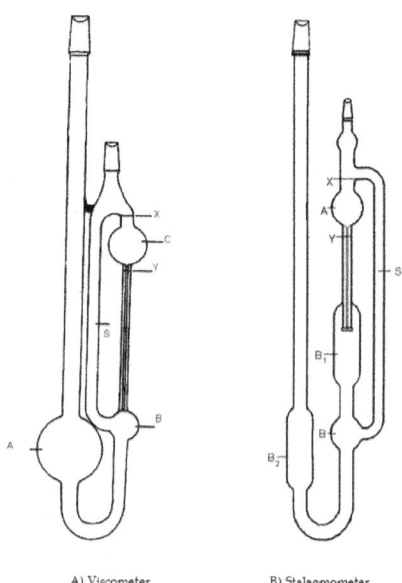

A) Viscometer   B) Stalagmometer

A VISCOMETER AND STALAGMOMETER

Mechanical Test

Macrorupture Test

The experimental samples after dimensions' ascertainment were subsequently subjected to tensile, modulus of rupture (flexural) and compressive tests in a VEB Thuringer Industrie Werk Rauenstein tensometer where a steadily increasing load would determine the rupture force of a crystal specimen until it shows least resistance. These tests were performed visually watching the fracture process. Though the tests are of conventional nature, yet can be related also to the macro scale, particularly relevant to engineering products.

Microdeformation Test

Metallic systems are suitable materials for micro hardness testing because of atomically rough surfaces. Indentations were induced on selected points chosen diagonally on an ansiotropically or isotropically grown specimen at room temperature ($\sim$300 K) using Vickers microhardness tester attached to an incident-light metallurgical research microscope (Neophot-2) in the applied load range 10×10^{-2} - 100×10^{-2} N. For each test, a very small diamond indenter having pyramidal geometry was forced into the surface of the specimen at room temperature and the size of the indent was found growing with rising applied load.

Microscopic Studies

Optical Microscopic Examination

The micro-slides of binary organic eutectic and akin mass composite phases grown anisotropically and isotropically were examined in a polarizing microscope to view the growth habits of crystallites which were later photographed.

Electron Microscope Examination

The metallic specimens grown isotropically and anisotropically were polished at room temperature by following the procedure referred in connection with the analogous problem (Caram et al., 1991). A thin layer of a specimen etched in ferric chloride was mounted on stub with gold coated holder and examined in a scanning electron microscope (Jeol T-330) for microgrowth observations. Several grown samples of eutectic and akin mass eutectic composites were viewed in this manner and the growth habits acquired by the growing eutectic phases during solidification process at different growth rates were photographed.

X-Ray Diffraction (Xrd) Studies

X-ray diffraction patterns of eutectics and non-eutectics were scanned in a Rigaku X-ray diffractometer with Cu K_α radiation of wave length 1.540 A^0 at room temperature to reveal their nature of combination.

OBSERVATIONS

The experimental evidences offer convincing similarity among binary organic systems in physical behavior and characterization. Metallic systems Sn-Cd and Pb-Sn also exhibit alike behavior with distinct morphology (regular) from that (complex regular) of Cd-Bi and Pb-Bi systems. In view of physical habits' resemblance of the systems, two representative examples, namely, naphthalene–o-nitrophenol from organic systems and Sn-Cd from metallic systems shall be the main components revealing the objectivity of the work. The melting temperatures of their constituent materials are recorded in Table1, also containing their entropies of fusion and α-values (where $\alpha = \xi \Delta S_f / R$, is called dimensionless entropy, since R is the molar gas constant) as well computed thereby.

Table 1: Melting temperatures, enthalpies and entropies of fusion, and α-values

material	melting temperature(K)	enthalpy of fusion (kJmol-1)	entropy of fusion (Jmol-1K-1)	α-value
naphthalene	353.00	19.00	53.81	6.47
o-nitrophenol	318.70	17.50	54.91	6.60
Sn	509.00	7.00	13.70	1.65
Cd	597.00	6.10	10.22	1.23

Diagrams of state representing the experimentally determined liquidus temperatures and theoretically computed ideal temperatures of naphthalene–o-nitrophenol and Sn-Cd systems are represented in Fig.1(a and b). The undercooling curve resulting in by experimentally determined temperatures of spontaneous crystallization for naphthalene–o-nitrophenol system is also included in its phase diagram (Fig.1a). Thermodynamic analysis of the systems reveals specific molecular interactions in their molten states and hence their non-ideal nature. The non-ideal character of the system is authenticated by excess functions €, namely, Gibbs free energy (G^E), entropy (S^E) and enthalpy

(H^E), computed by the following standard relations (Sharma et al.,2004b):

$$G^E = RT[x_1^l \ln\gamma_1^l + x_2^l \ln\gamma_2^l] \tag{1}$$

$$S^E = -R[x_1^l \ln\gamma_1^l + x_2^l \ln\gamma_2^l + Tx_1^l (\partial \ln\gamma_1^l / \partial T)_P + Tx_2^l (\partial \ln\gamma_2^l / \partial T)_P]$$

(2)

$$H^E = -RT^2[x_1^l \ln\gamma_1^l + x_2^l \ln\gamma_2^l] \tag{3}$$

where x_i^l and γ_i^l respectively, denote the mole fraction and activity coefficient of the eutectic phase i (i = 1, 2) in binary mixture of liquidus temperature T, with the enthalpy of fusion $\Delta_f H_i$, obtained by thermal analysis, at its melting point T_i . The superscript l refers to the condensed phase of the systems. The differential quantities ($\ln\gamma_1^l$ / T)$_P$ and ($\ln\gamma_2^l$ / T)$_P$ are the slopes of the liquidus lines obtained by plotting activity coefficients of the eutectic phases in eutectic mixtures versus their liquidus temperatures.

The quantitative magnitudes of G^E, S^E and H^E, obtained by Eqs. (1, 2 and 3) at different mole fraction compositions with their respective liquidus temperatures, are plotted in Fig.2. The minimal and maximal inflections respectively of G^E, and S^E and H^E at the eutectic compositions evidentially evince their obedience to the spontaneity criteria and Planck probability distribution (S = k ln w, k and w respectively being the Boltzmann constant and the configurational weight of the phase molecules). Microstructural parameters structuring the modal microstructures of the systems drive movement strength from excess functions (Sharma et al., 2008,2009) to overcome their faults called lamellar faults (Hunt & Jackson, 1966).

The following symmetry relations obtained from the limit of undercooling (Table 2) are found consistent over the whole of the compositional range expressing their obedience to the criteria of nucleation theory:

$$\xi = \frac{T_C}{T_m} \tag{4}$$

$$\frac{\Delta T}{T_m} = \frac{T_m - T_c}{T_m} \tag{5}$$

at which ΔT, T_m and T_c respectively, are the undercooling, liquidus and spontaneous crystallization temperatures. ξ is a crystallographic factor which is less than and almost one, and represents fraction of total number of neighbors situated in the newly formed crystal layer.

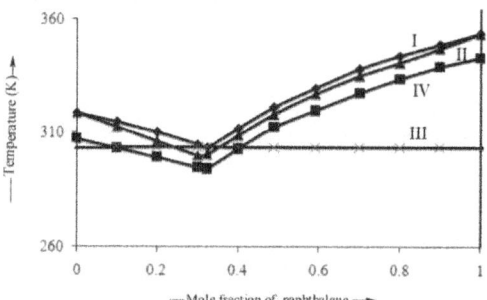

(a) Naphthalene– o-nitrophenol system
(I) liquidus temperatures curve
(II) ideal temperatures curve
(III) solidus temperatures curve
(IV) under cooling curve

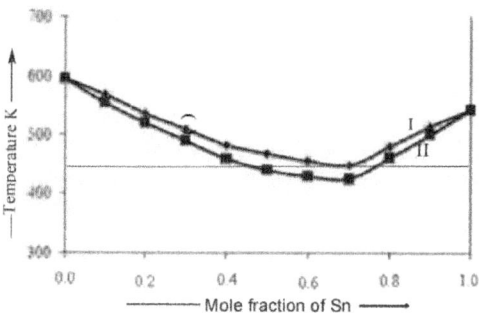

(b) Sn–Cd system
(I) liquidus temperatures curve
(II) ideal temperatures curve

Figure 1: Diagrams of State.

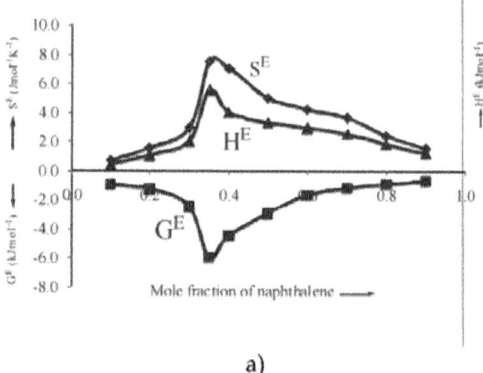

a)

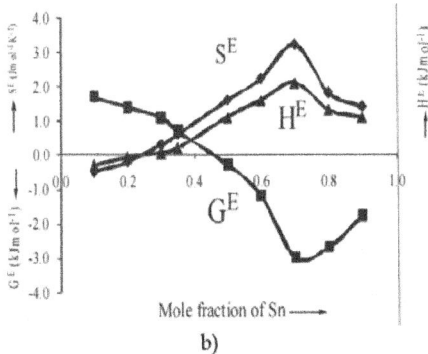

b)

Figure 2: Excess functions G^E, H^E and S^E, a) Naphthalene– o-nitrophenol system, b) Sn–Cd system.

Table 2: Heterogeneous nucleation data for the naphthalene-o-nitrophenol eutectic system. (e*: eutectic)

mole fraction of naphthalene	liquidus temperature Tm (K)	temperature of spontaneous crystallization Tc (K)	maximum undercooling ΔT=(Tm– Tc) (K)	ξ =Tc/Tm	ΔT/Tm x 10-2
0.0	318.70	307.20	11.50	0.96	3.61
0.1	315.20	303.10	12.10	0.97	3.84
0.2	311.70	299.00	12.65	0.96	4.06
0.3	304.70	294.70	10.00	0.97	3.28
0.33(e*):	303.70	293.20	10.50	0.97	3.46
0.4	313.70	304.70	9.00	0.97	2.87
0.5	321.20	312.40	8.80	0.97	2.74
0.6	329.70	319.80	9.90	0.97	3.00
0.7	339.90	327.70	12.20	0.97	3.59
0.8	343.20	333.70	9.50	0.97	2.77
0.9	348.70	339.10	9.60	0.97	2.75
1.0	353.50	343.70	9.80	0.97	2.77

The kinetics of anisotropic crystallization velocity, V, from the molten states of the eutectic and non-eutectic phases follows the parabolic law:

$$\log V = \log k + n \log \Delta T \qquad (6)$$

since k and n respectively, being the intercept and the slope are determined from the straight lines obtained by plotting log V versus log T (Fig.3) wherein n equals or almost equals 2 for the system. The observed anisotropic velocity of crystallization estimated from the eutectic melt is relatively low in comparison to its values from the non-eutectic melts (T = 10 K).

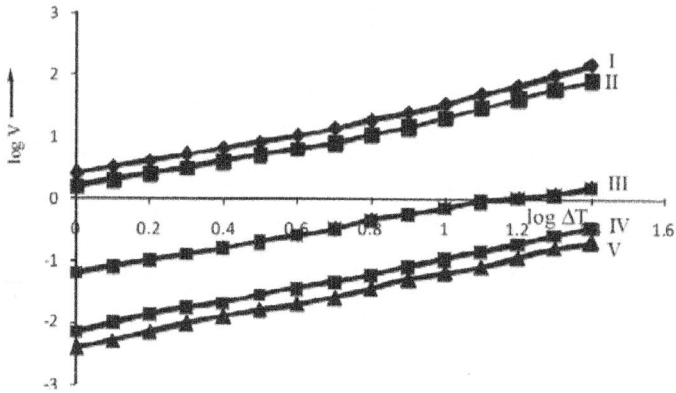

Figure 3: Anisotropic velocity of crystallization at various degrees of undercooling for naphthalene– o-nitrophenol system: (I) pure naphthalene; (II) pure o-nitrophenol; (III)-(V) 0.2, 0.4 and 0.3250 (eutectic) mole fractions of naphthalene.

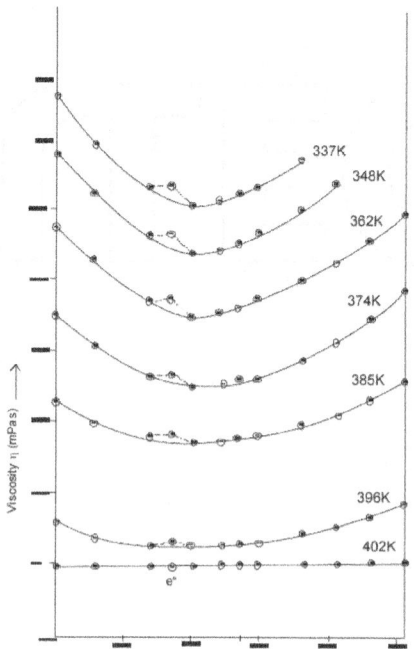

Figure 4: Viscosity-composition plot for naphthalene-o-nitrophenol system at different temperatures (e*eutectic).

The viscosities for different compositional melts of the naphthalene–o-nitrophenol system measured in the temperature range 337–402 K are represented in Fig.4, which evidentially predicts the anomalous viscous behavior by the eutectic compositional melt, particularly in the vicinal temperatures to its liquidus temperature and this unusual behavior starts declining with rising temperature and completely evanesces at 402 K.

The viscosity data of the system in the experimental temperature range follow the Arrhenius equation:

$$\log \eta = \log \eta_0 + \frac{Evis}{2.303RT}$$

(7)

the parameters and E_{vis} respectively, being the viscosity and activation energy for viscous flow of any compositional melt of the system at different rising temperatures T. The log values are plotted against $1/T$ for any compositional melt at different temperatures and the slope of the best straight line obtained thereby, determines the activation energy and the intercept of the line yields the constant $_0$ for that composition. The activation energy data (Table 3) for experimental compositional melts of the system obtained likewise reveal the predominance of activation energy for the eutectic melt over its values for the pre- and post-eutectic melts, which finally, consociates with a normal value of the system at 402 K (Fig.5).

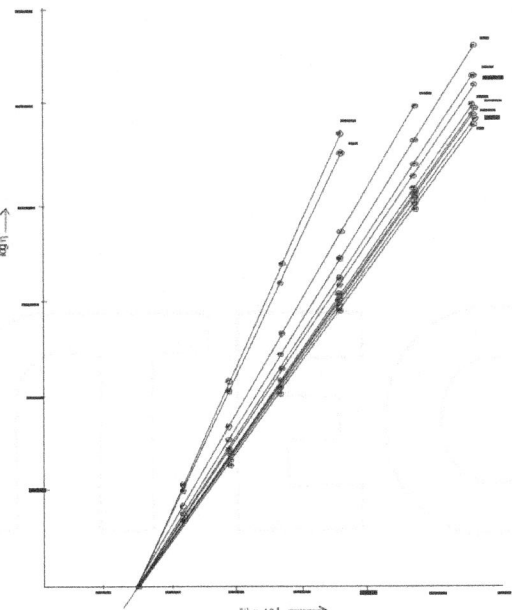

Figure 5: A plot of log η vs T for naphthalene-o-nitrophenol system.

I o-nitrophenol

XI naphthalene

II-X mixtures with 0.1, 0.26, 0.32 (eutectic), 0.38, 0.46, 0.52, 0.7, 0.8 and 0.9 mole fractions od naphthalene respectively

Table 3: Activation energy data in the temperature range 337-402 K for naphthalene-o-nitrophenol system in the paraffin thermostat with an accuracy of ±0.05 K

mole fraction of naphthalene	activation energy EVIS(X 10-2 kJmol-1)	mole fraction of naphthalene	activation energy EVIS(X 10-2 kJmol-1)
0.00	3.96	0.50	3.92
0.10	3.72	0.57	3.86
0.26	3.62	0.80	3.78
0.33 (e*)	3.88	0.90	3.64
0.38	3.64	1.00	3.56
0.46	3.82		

Alike, the surface tension, γ, measurements for different compositional melts of the system α-naphthylamine-diphenylamine in the temperature range 323–373 K (Fig.6) offers supporting evidence to the abnormal tendency of the phase molecules in the eutectic compositional melt.

The surface tension data of the system in the experimental temperature range obey Eotvos empirical equation:

$$\gamma = k(T_c - T) [d/M]^{3/2}$$

(8)

where k is the constant with numerical value 2.1 x 10^{-7} $JK^{-1}mol^{-2/3}$ valid for almost all materials and the parameters T_c, T, d and M respectively, are the critical temperature at which internal pressureequals 1 atmospheric pressure, experimental temperature, density and molar mass. The situation at which T_c equals T, the surface tension approaches zero. Figure 7 is a plot showing the variation of surface tension over its entire compositional

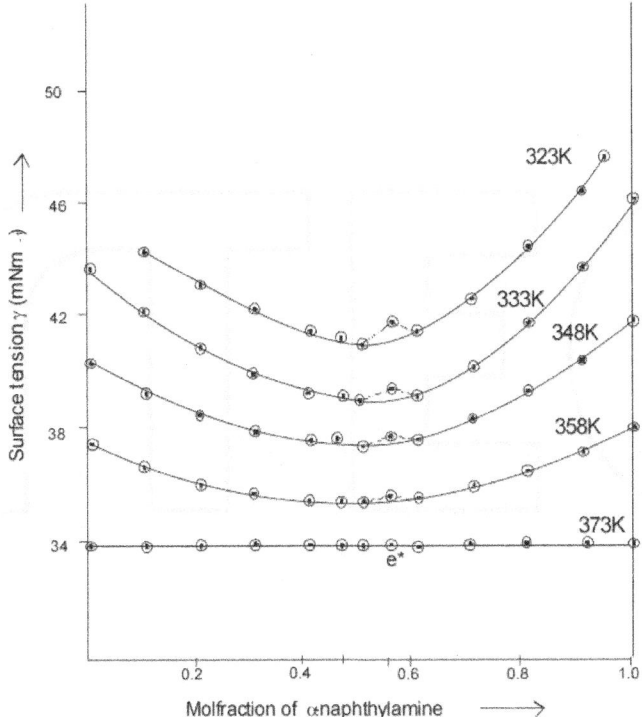

Figure 6: Surface tension-composition plot for -naphthalene-diphenylamine system at different temperatures, e* eutectic.

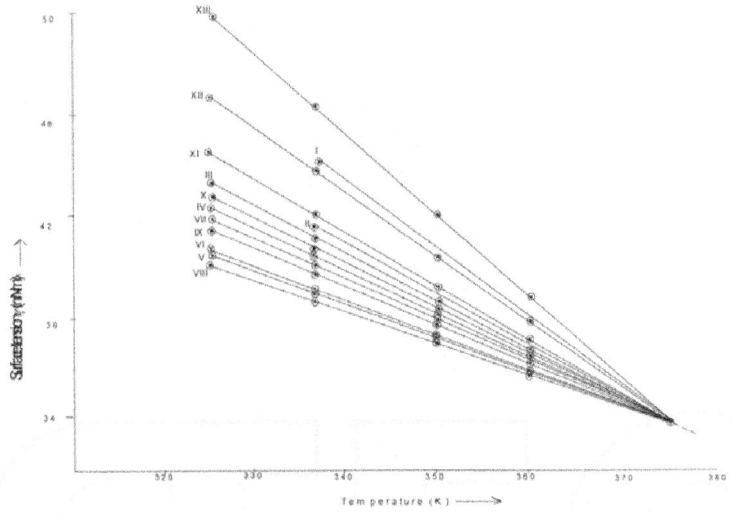

I diphenylamine

XIII α-naphthalene

II-XII mixtures with 0.1, 0.2, 0.3, 0.4, 0.46, 0.5, 0.55,(eutectic) 0.6, 0.7, 0.8, 0.9 molfractions α-

naphthalene respectively

Figure 7: A plot of γvs temperature for naphthalene-diphenylamine system range with increasing temperature. The temperature greatly effects the surface tension reducing it to its limiting value at $T = T_c$, an experimental observation possible only when there is no meniscus or barrier between the melt and the vapors (1 atm pressure). This implies that experimental observations of both viscosity and surface tension strongly support the excess functions predicting prominent molecular interactions and consequently, the structural changes in the eutectic melt evincing the endothermic nature of the systems.

The germinating crystallites from organic eutectic phases melt in an ice bath environment (~273 K) are of short size, aggressive, non-attaching and crossing each other showing no crystallite–matrix relationship resulting in overall morphology amply rich in small spherulites (Fig. 8a). The spherulitic density (spherulites per unit area) decreases with decreasing undercooling (Fig. 8b) and spherulitic size increases (~300 K) showing its centre as the seat of nucleation which finally evanesces (Fig.8c) with further undercooling decrease (~315 K). Nevertheless, the crystallites' growth habits of eutectic and akin mass eutectic compositions can be transformed from spheric or irregular morphology to unique fibrous and lamellar structures consisting of crystallites which are non-aggressive, attaching and embedded parallel to each other reinforcing the matrix by anisotropic growth modes of solidification.

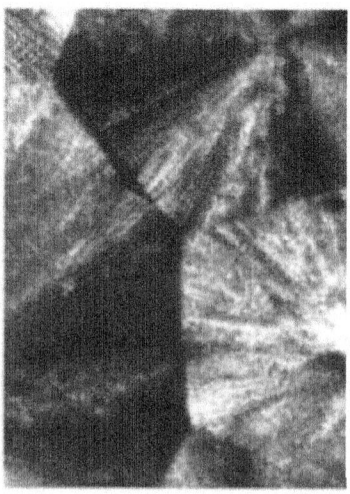

(a)

(b)

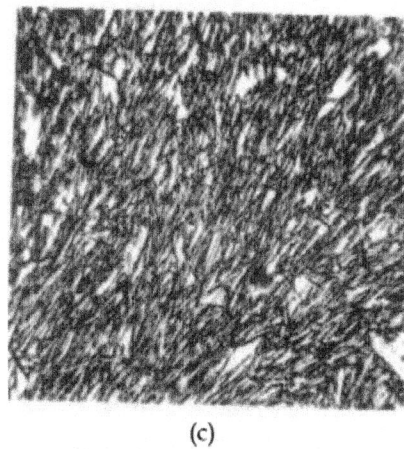

(c)

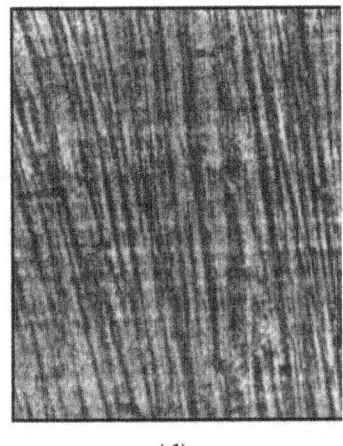

(d)

(e)

(f)

Figure 8: Morphological diversity of naphthalene–o-nitrophenol eutectic composite materials by different modes of solidification (100 x). Microstructure of eutectic composite material in (a) ice bath (~273 K), isotropic growth, (b) at room temperature (~300 K), random growth, (c) at 315 K, (d) at moderate anisotropic growth from bottom to top (~2.90 X 10^{-7} m^3s^{-1}), (e) akin mass eutectic composite material at moderate anisotropic growth from bottom to top (~2.90 X 10^{-7} m^3s^{-1}), (f) eutectic composite material at anisotropic growth from bottom to top (~2.14 X 10^{-7} m^3s^{-1}).

In the domain of solidification rates, microstructures represented by Figs. 8 d and 8e are respectively discovered for eutectic and akin mass eutectic composites of naphthalene–o-nitrophenol system at moderate anisotropic growth rate (~2.90 X 10^{-7} m^3s^{-1}). Both the microstructures indispensably exhibit long and continuous crystallites with damaged free surfaces embedded parallel to each other along the growth direction. The akin mass eutectic composite structure is annexed superior to that of eutectic composite, vis-à-vis strength point of view. Figure 8f is a microstructure of the eutectic phases developed at anisotropic growth velocity ~2.14 X 10^{-7} m^3s^{-1} showing crystallites of short size and disconnected with aligning tendency to the longitudinal growth direction. On the contrary, pure eutectic components grew out from their respective melts as lamellar cells that crystallized either from the bulk of the melt or through secondary nucleation during the entire anisotropic solidification process. In view of anisotropic growth ability to organize the growing eutectic phases from the molten state, microstructures represented by Fig.9 (a, b and c) comprising of aligning lamellae, respectively are the specimens of Sn-Cd eutectic phases grown at moderate (~2.90 X 10^{-7} m^3s^{-1}), post-moderate (~3.50 X 10^{-7} m^3s^{-1}) and pre-moderate (~2.30 X 10^{-7} m^3s^{-1}) anisotropic growth rates. The basic distinction in morphologies of non-metallic and metallic eutectic systems lies in the thermodynamic concept (Ovsienko et al., 1980) that coupled growth occurs when both phases have low entropies of fusion ($\alpha<2$); i.e.; the lamellae or crystallites of the phases grow in contact resulting in regular structure with their round growth fronts and the overall growth is termed non-faceted–non-faceted (nf-nf). A coupled growth doesn't occur when both the eutectic phases have high entropies of fusion ($\alpha>2$) and consequently a regular structure result in consisting of crystallites or lamellae of the phases that grow side by side near each other with sharp growth fronts and the type of the growth is differentiated faceted-faceted (f-f). A complex regular structure comprising of either nf- f lamellae or nf-f crystallites is produced when the entropy of fusion is low ($\alpha<2$) in one phase but high ($\alpha>2$) in the other and the growth being the intermediate case, is distinguished non-faceted–faceted (nf-f). In the naphthalene–o-nitrophenol system, both the phases have high entropy of fusion ($\alpha>2$) furnishing an irregular structure, whereas in Sn-Cd system, both the phases have low entropy of fusion producing regular morphology. Regular morphology implies that the spacing among lamellae or crystallites must appear constant throughout the structure; otherwise irregular in nature. For this very reason, rod-type growth has a perfect ordered structure merely because of periodic constant inter-rod spacing, whereas lamellar structure proves to be less regular sheerly because of the movement of lamellae or crystallites termed lamellar or fiber faults that would become the cause of slight uncertainty in their spacing consistency.

Macro-rupture of naphthalene–o-nitrophenol system is virtually a collection of the following stress-strain fractures (Boyer, 1999; Callister,2006 & Kollmann, 1975):

- Modulus of rupture, $Y_{rup} = \dfrac{PL}{\pi r^3}$ (circular cross section)

$$(9)$$

- Tensile strength, $T_{rup} = \dfrac{P}{\pi r^2}$

$$(10)$$

- Compressive strength, $\sigma_{rup} = \dfrac{P}{\pi r^2}$

$$(11)$$

where the parameters P, L and r respectively, are the applied load in Newton force (N), span, and radius of an experimental specimen in metres, herewith P and r being dependent on the dimensions of the specimen and nature of the fracture, have variable magnitudes. In each fracture, the specimens yielding strength deviations by >15% from the averaged result were not included in the computed mechanical property data presented for eutectic and akin mass eutectic composite phases grown at both moderate anisotropic (2.90 X 10^{-7} m^3s^{-1}) and isotropic (~273 K) solidification rates, in Table 4.

(a)

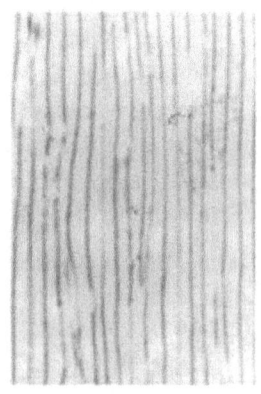

(b)

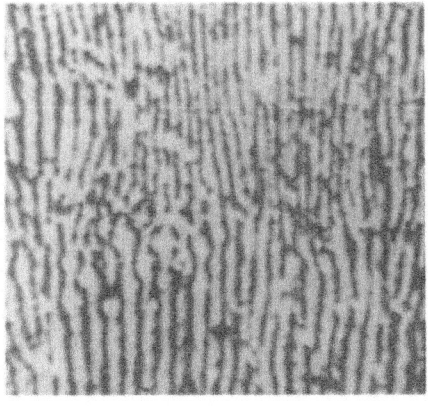

(c)

Figure 9: Lamellar growth of Sn-Cd eutectic alloy in the moderate growth region (1500x): (a) lamellar Sn-Cd eutectic at moderate anisotropic growth from bottom to top (~2.90 X 10-7 m3s⁻¹), (b) disconnected lamellar Sn-Cd eutectic at anisotropic growth from bottom to top (~3.50 X 10-7 m3s⁻¹),(c) disconnected and short lamellae structure of Sn-Cd eutectic at anisotropic growth from bottom to top (~ 2.30 X 10-7 m3s-¹).

The mechanical observations data, obtained by Eqs. (8, 9 and 10) for the entire composition range of the naphthalene–o-nitrophenol system at constant moderate anisotropic growth (~2.90 X 10^{-7} m^3s^{-1}), are plotted in Fig.10, implicitly indicating the predominance of the akin mass eutectic composition. In search of authenticity onto this experimental evidence, both eutectic and akin mass eutectic composite materials grown at constant but different anisotropic growth rates, were subsequently subjected to tensile, flexural (modulus of rupture), and compressive tests and the computed tensile strength data are represented in Fig.11, while data for modulus of rupture and compressive modes are recorded in Table 5, supporting the predominance of the akin mass eutectic composite over eutectic composite, both expressing obedience to the Weibull distribution. Evidently, this aspect generates the strength-growth relationship which follows an identical form of the Weibull probability distribution curve inculcating the obedience of the microstructural paramaters, namely, crystallite diameter, crystallite length, crystallite length distribution, volume fraction of crystallites, and the alignment and packing arrangements of crystallites, to the distribution.

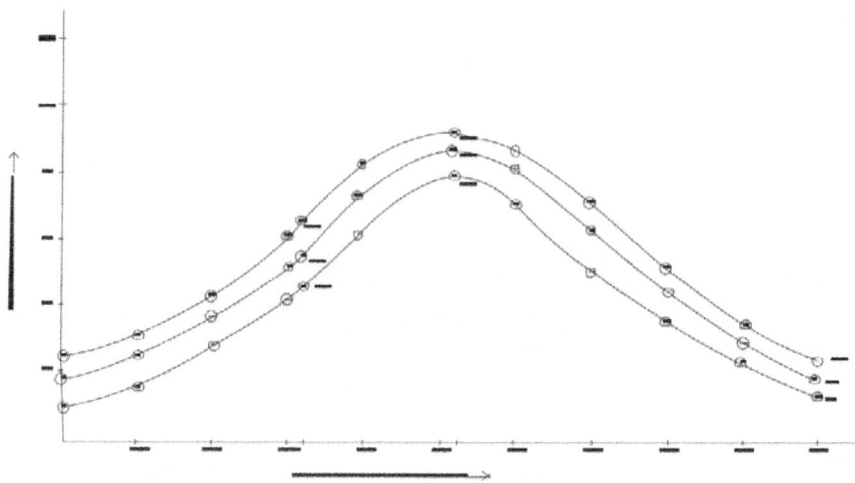

Figure 10: Variation of mechanical properties of naphthalene-o-nitrophenol system at moderate growth ($2.90 \times 10^{-7} m^3 s^{-1}$) (i) modulus of rupture (ii) tensile strength (iii) compressive strength a* akin mass composition, e* eutectic.

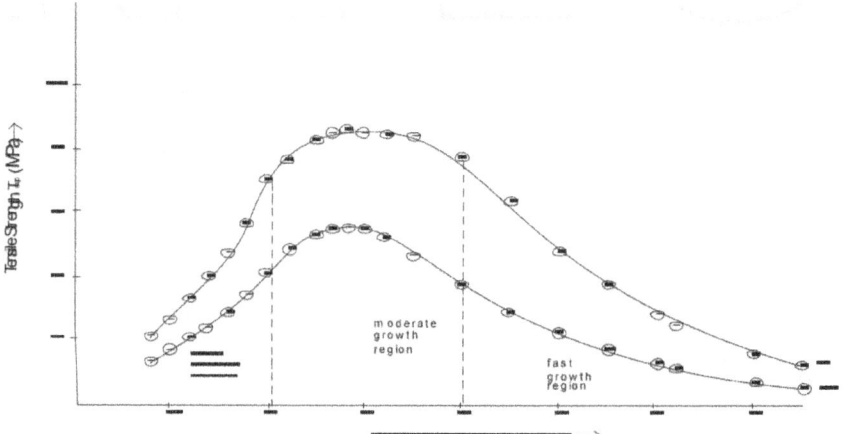

Figure 11: Plots of tensile strength at variable growth (i) akin mass composition and (ii) eutectic composition of the naphthalene-o-nitrophenol system.

Table 4: Mechanical properties of akin mass and eutectic compositions of naphthalene-o-nitrophenol system at moderate anisotropic (~ 2.90 x10^{-7}m^{3}s^{-1}) and isotropic (~273 K) growth rates

Mechanical property	akin mass composition		Eutectic composition		Noneutectic materials			
			naphthalene		o-nitrophenol			
	moderate anisotropic growth	isotropic growth	moderate anisotropic growth	isotropic growth	moderate anisotropic growth	isotropic growth	moderate anisotropic growth	isotropic growth
(i) Modulus of rupture Yrup (Mpa)	7.90	2.96	5.10	2.50	1.30	0.60	1.00	0.50
(ii) Tensile Strength Trup (Mpa)	8.60	3.40	6.00	2.90	1.80	1.0	1.80	1.00
(iii) Compressive Strength σ rup (Mpa)	9.20	3.70	6.70	3.30	2.30	1.10	2.40	1.30

Table 5: Variation of modulus of rupture and compressive strength with growth velocity of naphthalene- o-nitrophenol binary system

growth velocity (V X 10-7m3s-1)	akin mass composition		eutectic composition	
	modulus of rupture Yrup (MPa)	compressive strength σrup (MPa)	modulus of rupture Yrup (MPa)	compressive strength σrup (MPa)
0.8	1.10	2.80	0.70	1.50
1.0	1.80	3.50	1.00	1.90
1.2	2.80	4.40	1.30	2.40
1.4	4.20	5.10	1.70	2.70
1.6	5.40	6.00	2.30	3.70
1.8	6.20	6.80	2.80	4.30
2.0	6.80	7.60	3.30	5.00
2.2	7.10	8.20	3.80	5.70
2.5	7.70	8.90	4.20	6.30
2.7	7.80	9.10	4.40	6.50
2.8	7.90	9.20	4.50	6.60
3.0	7.80	9.10	4.40	6.50
3.2	7.60	9.00	4.30	6.40
3.5	7.10	8.80	3.80	6.00
4.0	6.10	7.80	2.70	5.20
4.5	4.80	6.20	2.00	3.80
5.0	3.70	4.70	1.50	2.90
5.5	2.60	3.80	1.10	2.30
6.0	2.10	3.00	0.80	1.80
6.2	1.70	2.70	0.60	1.60
7.0	1.20	1.80	0.30	1.50
7.5	0.80	1.30	0.20	0.80

Micro deformation test virtually represents the Vickers microhardness, H_v, which is a measure of a materials resistance to localized plastic deformation (e.g., a small dent or a scratch) by the stress-strain relation (Beigh et al., 1995 & Hayden, 1965):

$$H_v = 1.8544 \, P/d^2$$

(12)

herewith, P signifies the same as mentioned above, whereas the parameter d denotes the average diagonal length of the indentation mark in meter and varies with the size of the cracks.

The variation of Vickers microhardness for specimens of the eutectic alloy Sn-Cd grown by different modes of experimental anisotropic solidification rates at constant applied load of 50×10^{-2} N for a specified interval of 20s, offers supporting evidence to the essence of an identical form of the Weibull distribution curve (Fig.12) exhorting the strength-growth relationship. The specimens of the eutectic alloy Sn-Cd obtained by moderate anisotropic ($\sim$2.90 $\times$ 10^{-7} $m^3 s^{-1}$) growth process inflicted with indentation impressions by a constant load (60×10^{-2} N) at different indentation times (Fig.13 a, b and c), measurement of crack length and growing indents' dimensions by a variable load are illustrated in terms of photomicrographs in Fig.13 (d, e and f). A maximum care was exercised to consider only well defined cracks developed during indentation process. The average crack length of all such cracks was estimated for a particular indentation impression. The crack length was measured from the centre of the indentation mark to the tip of the crack (Fig.13d). The size growth with a variable load is represented in Fig.13 (e and f).

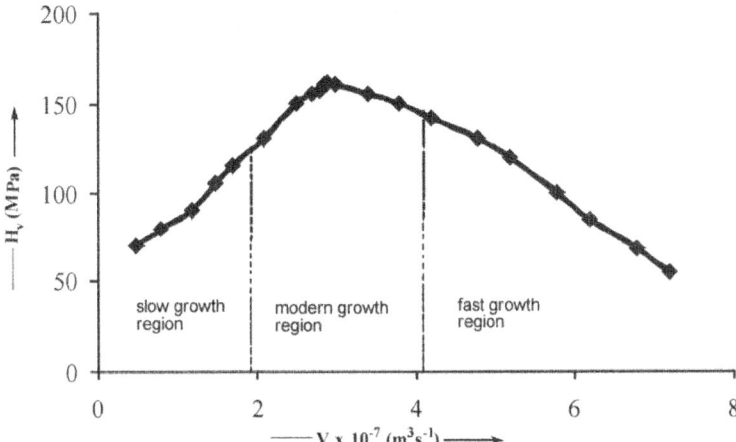

Figure 12: Variation of micro hardness with growth velocity for the eutectic alloy Sn-Cd at constant applied load of 50×10^{-2} N for a specified interval of 20 s.

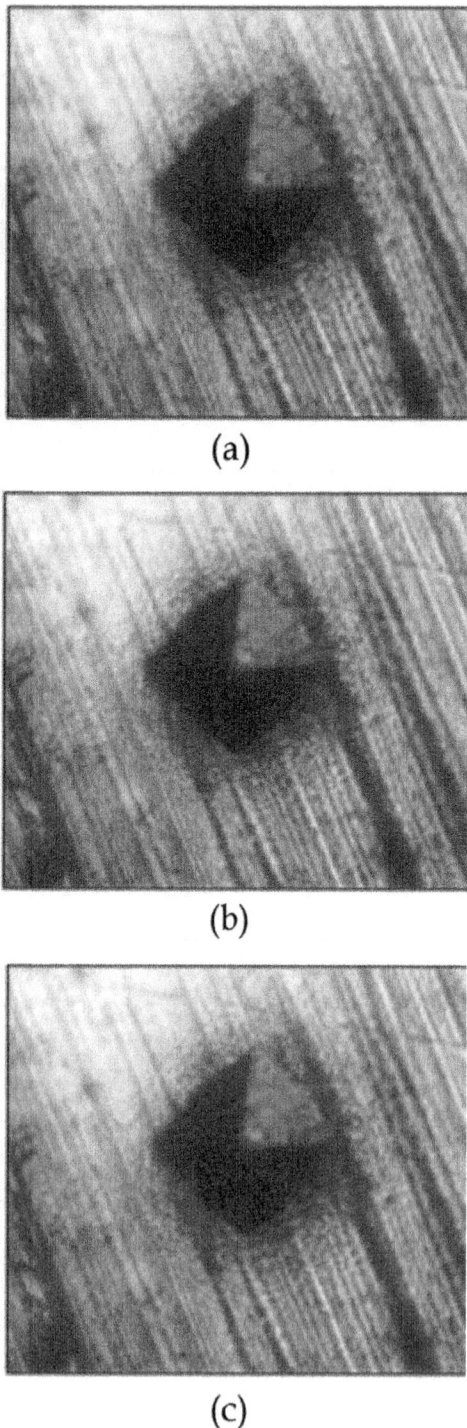

(a)

(b)

(c)

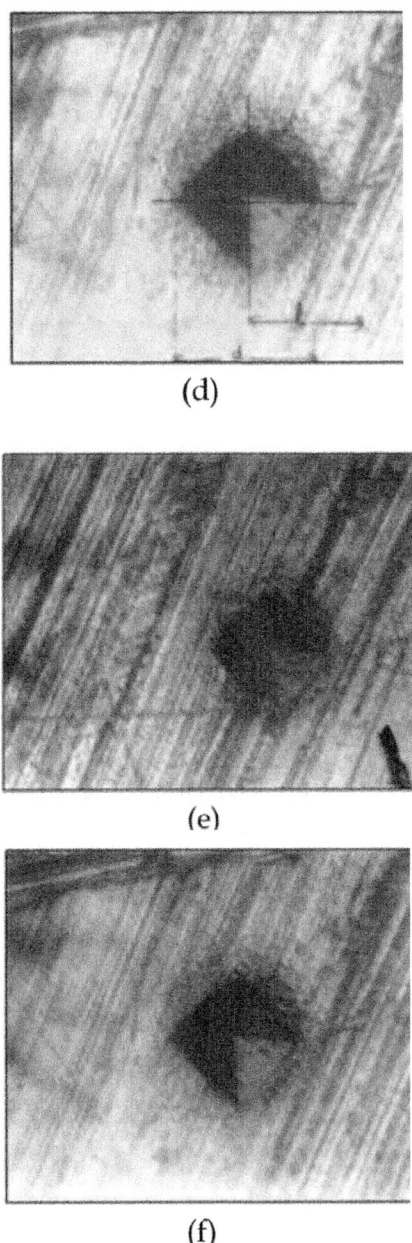

(d)

(e)

(f)

Figure 13: Photomicrographs showing indentation impressions inflicted on the aniso-tropic specimens of the eutectic alloy Sn-Cd (625x): (a) size of indentation with a load of 60×10^{-2}N at 50s; (b) size of indentation with a load of 60×10^{-2}N at 100s;(c) size of indentation with a load of 60×10^{-2}N at 150s; (d) measurement of crack length;(e) size of indentation with a load of 20×10^{-2}N; (f) size of indentation with a load of 40×10^{-2}N.

Figures 14 and 15 respectively predict the variation of microhardness with increasing indentation time and applied load. The analysis of microhardness is summarized in Figs. 16-19.

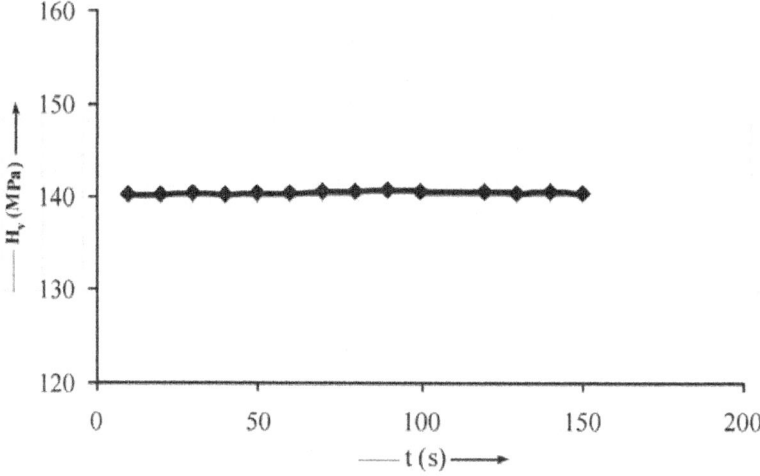

Figure 14: A plot showing invariability of micro-hardness with variable indetation time for eutectic alloy Sn-Cd for constant applied load. P $(30 \times 10^{-2}$ N$)$.

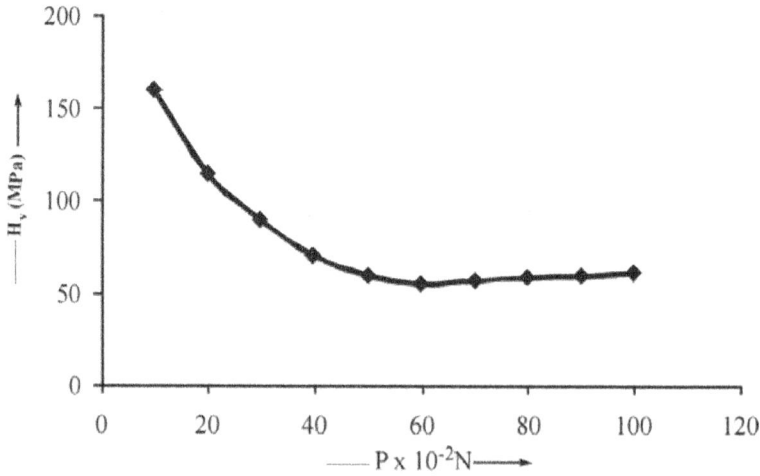

Figure 15: Dependence of micro-hardness on applied load for eutectic alloy Sn-Cd for constant indentation time, t (20 s).

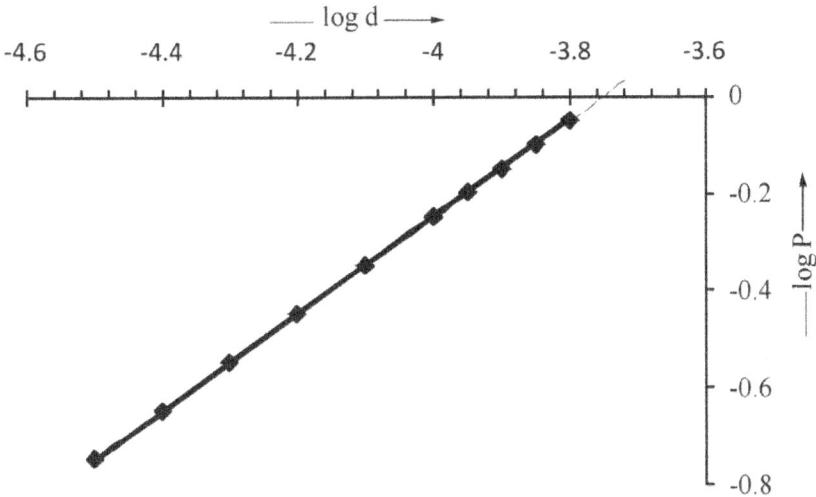

Figure 16: Relationship between log P and log d for the eutectic alloy Sn-Cd.

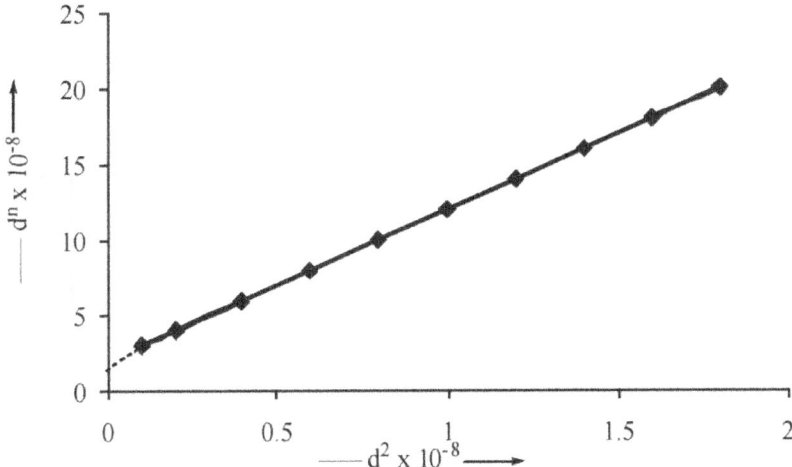

Figure 17: Variation of d^{π} and d^2 for the eutectic alloy Sn-Cd.

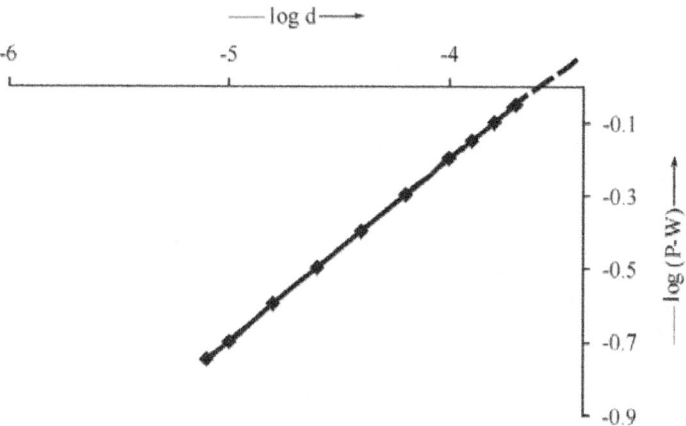

Figure 18: Relationship between log (P-W) and log d for eutectic Sn-Cd.

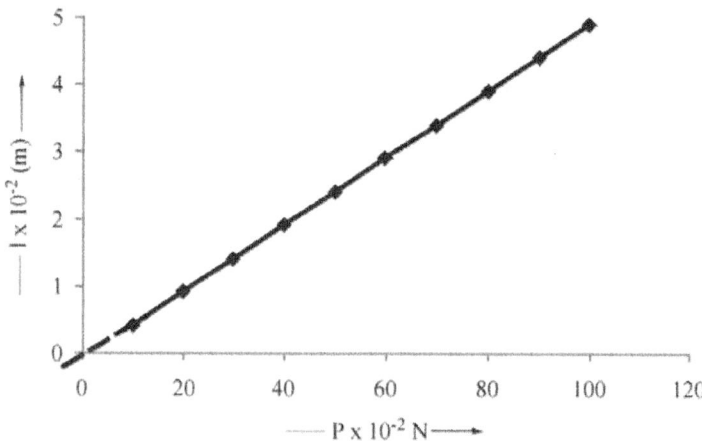

Figure 19: A plot showing variation of cracklength (l) with applied load (P) for eutectic alloy Sn-Cd.

The parameters extracted from the analysis are recorded in Tables 6-7.

Table 6: Hardness analysis constants for the eutectic alloy Sn-Cd

eutectic alloy	n	$10^6 Nm^{-2}$ k1	$10^6 Nm^{-2}$ k2	k2/k1	W (10^{-3}N)	W/k1 ($10^{-8}m^2$)
Sn-Cd	1.5672	1.3824	7.4155	5.3642	8.7868	6.3562

Table 7: Hardness parameters for the eutectic alloy Sn-Cd

applied load P x 10-2N	crack length (l x 10-2m)	**fracture toughness Kc (Nm-3/2)**	**brittleness Bi (106m-1/2)**	Hardness Hv (MPa)	yield strength y (MPa)
10	0.3	89.29	1.79	160	53.33
20	0.7	84.03	1.37	115	38.33
30	0.9	50.19	1.79	90	30.00
40	1.2	43.62	1.60	70	23.33
50	1.6	35.36	1.69	60	20.00
60	1.9	32.72	1.68	55	18.33
70	2.1	32.89	1.73	57	19.00
80	2.7	25.74	2.29	59	19.67
90	3.2	22.48	2.67	60	20.00
100	3.4	22.78	2.68	61	20.33

The XRD data summarized in Table 8 and X-ray patterns represented by Fig. 20 for the eutectic alloy Pb-Bi, selected as representative example in the current work, specify that the number of peaks and intensities exhibited by the eutectic alloy are complementary to its constituent phases.

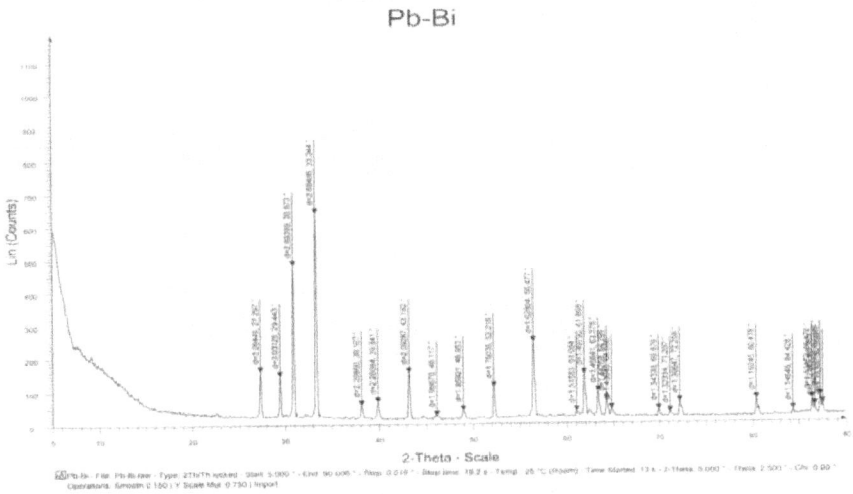

Figure 20: X-ray diffraction patterns of the eutectic alloy Pb-Bi

Table 8: X-ray analysis for the eutectic alloy Pb-Bi

S. No	Pb Pos [2(TH]	Pb d-spacing [Å]	Bi Pos [2(TH]	Bi d-spacing [A(]	Pb-Bi eutectic Pos [2(TH]	Pb-Bi eutectic d-spacing [A]	Corresponding Planes (hkl)
1	6.487	13.6113	6.224	14.189	22.469	3.95389	Bi (003)
2	28.918	3.0850	22.220	3.997	27.297	3.26449	Bi (012)
3	31.448	2.8423	22.805	3.8962	29.443	3.03125	Pb (110)
4	36.440	2.4636	24.091	3.6911	30.873	2.89399	Pb (111)
5	43.381	2.0842	26.821	3.3213	33.344	2.68499	Pb (111)
6	49.973	1.8236	27.427	3.2492	38.107	2.35960	Bi (104)
7	52.407	1.7444	31.502	2.8376	39.841	2.26084	Bi (110)
8	56.271	1.6335	37.740	2.3817	43.192	2.09287	Pb (200)
9	62.314	1.4888	38.211	2.3524	46.117	1.96670	Bi (113)
10	65.414	1.4255	39.872	2.2589	48.953	1.85921	Bi (202)
11	77.135	1.2353	44.826	2.0202	52.218	1.75035	Pb(220)
12	77.397	1.2321	45.684	1.9843	56.477	1.62804	Bi (024)
13	85.579	1.1339	46.199	1.9634	61.084	1.51583	Bi (200)
14	85.852	1.1310	48.511	1.8751	61.895	1.49790	Pb (311)
15	88.363	1.1052	48.954	1.8591	63.376	1.46641	Bi (211)
16	88.646	1.1024	49.561	1.8378	64.299	1.44758	Bi (122)
17	89.714	1.0921	56.286	1.6331	64.856	1.43648	Pb (222)
18			59.566	1.5507	69.979	1.34333	Bi (214)
19			61.325	1.5140	71.207	1.32314	Bi (009)
20			62.434	1.4862	72.258	1.30647	Bi (300)
21			64.761	1.4383	80.478	1.19245	Bi (208)
22			67.682	1.3832	84.428	1.14645	Bi (119)
23			71.013	1.3262	86.472	1.12452	Bi (101)
24			72.110	1.3087	86.763	1.12149	Bi (217)
25			73.917	1.2812	87.328	1.11568	Bi (217)
26			75.542	1.2576	87.590	1.11302	Pb (420)

ANALYSIS AND DISCUSSION

The deviations of the systems from ideal regions covered by their respective diagrams of state (Fig.1) consisting of solidus-liquidus equilibrium curves, predict specific interactions between unlike phase molecules or unlike atoms in the miscibility phenomenon. The thermodynamic analysis of the systems confirms the deviations (Fig.2) and comprehensively yields quantitative idea of the interactions, inasmuch the excess functions are being computed from experimentally determined parameters, are reliable and discovered expedient for predicting the phase equilibrium curves of binary systems (Sharma et al.,2004).

The heterogeneous nucleation data obtained from the limit of undercooling (Table 2) provide a lucid consistency in the symmetry relations, $\xi = T_c/T_m$ and $\Delta T/T_m$, which certainly authenticates the nucleation theory for the system entirety. This essential constancy in symmetry relations implicitly explores a definite relationship existing between the nucleating solidus form obtained from the limit of undercooling and the liquid structure conceived by the entropy of fusion.

The critical analysis of the plot (Fig. 3) shows that the anisotropic velocity of crystallization from binary compositional melts essentially decreases significantly with the gradual addition of one eutectic phase to the other. Consequently, the inferential interference of the eutectic phases decreases the crystallization velocity from the eutectic melt ($10 \text{ X } 10^{-4} \text{ms}^{-1}$) which is much lower than its value from the non eutectic melts; naphthalene ($158 \text{ X} 10^{-4} \text{ms}^{-1}$) and o-nitrophenol ($92 \text{ X} 10^{-4} \text{ms}^{-1}$) for fixed undercooling $\Delta T = 10K$. The kinetics of the anisotropic crystal growth being dependent on parabolic form, follow the dislocation mechanism wherein eutectic phases would diffuse and grow at the sites at which surface with the stages is formed owing to lattice imperfections and screw dislocations. These imperfections do intersect the surfaces and produce steps of one or more molecular diameters in height. These steps are the center of lattice disturbance and wind themselves in spirals during crystal growth. The logical growth concept extracted from the theoretical interpretation of the experimental observations is consistent with the diffusion growth of the eutectic phases from their binary melt that the growing phases mutually sustain the diffusion process through melt layer becoming rich in their concentration at different times of successive nucleation.

The anomalous behavior of viscosity (Fig.4) and surface tension (Fig.6) for the eutectic melt furnishes supporting evidence to the essence of molecular interactions conceiving magnitudes already delivered by the excess functions. The most probable structural change (configurational weight) in phase molecules occurs in the eutectic compositional melt owing to three phases' coexistence, resulting in significant molecular clusters' formation rich in one of the eutectic phases. The degree of molecular clustering tendency decreases as the temperature is gradually raised that would finally become the cause of limiting values of rheological parameters. The explanation for the higher value of the activation energy for the eutectic melt may be attributed to the following contributions: (i) the activation energy for viscous flow; (ii) the energy essentially required overcoming the molecular clusters' tendency. The molecular clustered array of atoms would inevitably result in an increase both in viscosity and surface tension, and a reduction in surface area. Consequent thereupon, the thermal energy expended to break the clusters would certainly exceed activation energy. An increasing temperature vitiates clustering phenomenon and the particular temperature whereat it completely evanesces only the first contribution will be the activation energy. In the current work, the cluster-free flow is evidentially illustrated for viscosity at 402K in Fig.4 and for surface tension at 373 K (Fig.6). Likewise, the activation energy for eutectic compositional melts also consociates with normal value of the systems as indicated by dotted lines in Fig.5 for viscous flow of naphthalene–o-nitrophenol system at 402K, and in Fig 7 for loose surface flow of α-naphthylamine-diphenylamine system at 373K.

An irregular morphology (Fig 8 a –c) by f-f growth of eutectic phases from molten state results in because of their high entropies of fusion, since the formation of packs of their crystallites from the eutectic melt would arise under the influence of large thermal and mechanical stresses during the growth process. These stresses seem to be the cause of splitting the main single crystallite into separate single crystallites or groups of single crystallites in the absence of orientation relationship among faceted crystallites, eventually accomplishing an irregular morphology. The crystallite spherulitic form occurs at a large undercooling (~273 K) merely because the rich eutectic phase first nucleates and grows depleting the vicinal melt to be rich in other phase that also nucleates at a certain supersaturation. Alike, both the phases grow side by side. The large thermal and mechanical stresses influence the splitting of crystallites to occur along the direction and successive splitting lead to the formation of spherulites. As mentioned in the observation section, the moderate anisotropic growth is predominant finding in the domain of anisotropic solidification rates because of its aesthetics ability to produce in- situ oriented composite materials consisting of crystallites (Fig.8 d–e) and lamellae (Fig.9a) which are attaching, nonaggressive and embedded parallel to each other exhibiting nearly consistent periodic orientation relationship. Microstructure (Fig.9b) consisting of long and continuous parallel lamellae indicates lamellar faults (Hunt &Jackson, 1966). Microstructures (Figs.8 f and 9 c) consisting of crystallites or lamellae of short size, exhibit their disconnected but non aggressive alignment along the growth direction. The supercrescent crystallites or lamellae growth virtually produces the complete lamellar or rod –type microstructures.The eutectic phases with α >2 grow side by side near each other from binary melt resulting in lamellar structures of long and continuous non aggressive crystallites lamina (Fig.8d-e), whereas eutectic phases with α <2 grow supercrescently (edge-wise) from the melt to produce lamellar or rod-type lamina consisting of long and continuous nonaggressive lamellae which are embedded parallel to each other showing orientation relationship almost consistent throughout the structure (Fig. 9a).

An insight analysis of Table 4 reveals approximately one and a half to two fold enhancement in each mode of mechanical property of the akin mass eutectic composite material at the moderate anisotropic growth velocity (~2.90 X 10^{-7} m^3s^{-1}) in comparison to the eutectic composite material and many fold superiority over its isotropic growth (~273 K) and constituent materials whether grown anisotropically or isotropically. The plot of Fig.10 drawn between different modes of macrohardness and the entire range of composition of the naphthalene-o-nitrophenol system at nearly constant moderate anisotropic growth velocity (~2.90 X 10^{-7} m^3s^{-1}) strengthens the quantitative analysis by showing the predominance of the akin mass eutectic composition. Consequently, the curve (Fig.11) divides the experimental range of anisotropic growth velocity into three regions, namely, (i) slow growth region; (ii) moderate growth region and (iii) fast growth region.Among which, the moderate growth region, as is evident from the plot (Fig.11), would seem to be the most probable for producing lamellar or rod-type structures of both eutectic (Fig.8d) and akin mass eutectic (Fig.8e) composite materials, since microstructural parameters appear to be nearly obeying the Gauss distribution. Consequentiality, the microstructural crystallites of the akin mass composite material nearly follow the rod- type growth habits (Fig.8e) reinforcing the matrix where there is a perfect crystallite- matrix bond. On the contrary, the crystallites of the eutectic microstructure (Fig.8d) emerge with lamellar growth habits reinforcing the matrix but the crystallite- matrix bond exist with less perfection.The mechanical parameters of both the composite materials in the slow and fast growth regions are found comparable (Table 5 and Fig.11) because microstructural parameters express their obedience to the Weibull distribution in these regions showing no crystallite matrix relationship (Fig.8 a - c).The analysis necessarily involves a physical understanding of the relationship between the microstructures of materials and their mechanical properties.Evidentially, the curve (Fig.11) has two cut-off points corresponding to a lower strength limit in the slow and fast growth regions, and an upper strength limit in the moderate growth region. The later is equivalent to the theoretical strength of the matrix reinforced crystallites with damaged free surfaces, practically having no density of dislocations; i.e.; internal defects or surface flaws responsible for reduced strength. Likewise, the microhardness of the eutectic alloy Sn-Cd (Fig. 12) can be explained essentially involving its microstructures presented inFig.9 on the similar dependence pattern with variable anisotropic growth velocity as tensile strength for the eutectic and akin mass eutectic composite materials of the naphthalene-o-nitrophenol system, follows in Fig.11. Implicit in the plots (Figs.11 and 12) is the concept that the variation of an anisotropic mechanical property over the entire range of growth velocity furnishes an evidence of its dependence as linear, optimum, and linear respectively in slow, moderate, and fast growth regions of solidification. An interesting finding to be noted from Tables 4 - 5is that, both eutectic and akin mass eutectic composite materials with isotropic growth (~273 K) are respectively found nearly twice and thrice stronger than that of their constituent materials merely because of their high specific modulus and high specific strength due to alignment of the crystallites although having dislocations. Likewise, moderate anisotropic growth of pure constituent materials is superior to their isotropic growth. Besides, the quantitative perusal of mechanical parameters presented in Tables 4 and 5,explores that the compressive mode slightly exceeds the tensile mode, which in turn shows an edge on modulus of rupture ($_{rup} > T_{rup} > Y_{rup}$).The inequality existing between mechanical modes essentially involves crystallite's length, which in the current work, accomplishes with the attachment of alike crystallites at different times and their efficiency in gripping and the reinforcing the matrix decreases with the reduction of crystallite's length. Crystallite-ends play important role in the fracture of short crystallite composite materials (Fig.8f and Fig.9c) and also in continuous crystallite composite materials (Fig.8d and Fig.9b), since the long crystallites may break down into discrete lengths.

Figure 13 a-c reveals that the indentation size by a constant load (60×10^{-2}N) is independent of variable indentation times inferring the microhardness (H_v) to be so. The observation is strongly supported byFig.14 indicating practically no change in the microhardness with the variable indentation time for a constant applied load of (30×10^{-2}N) at room temperature, which is consistent with the plastic deformation of the eutectic alloy that remains unaffected with variable indentation times. The nonlinear variation of microhardness (Fig.15) with variable applied load in the range $10 \times 10^{-2} - 100 \times 10^{-2}$ N at constant indentation time, t (20s) shows that H_v decreases with increasing load until about 60×10^{-2}N and thereafter, H_v tends to saturation which is full at 70×10^{-2}N. The qualitative explanation of the variation follows that the indenter penetrates only surface layers at small loads, consequently, the effect is more pronounced at these loads. However, the increasing load increases the penetration depth that damages the inner layers more effectively to shattering the ability of hardness and ultimately, H_v appears to be independent of further applied load. The variation is in good agreement with the microhardness increase during early stages of plastic deformation (Brookes, 1986) but is contrary to Kick's law that H_v remains constant irrespective of the magnitude of applied load, P (Ascheron et al.,1989),qualitatively:

$$P = k_1 d^n \tag{13}$$

the Meyer's index, n = 2, accounts for constant H_v, and k_1 is a constant. A linear plot between log P and log d (Fig.16) of Eq. (13) yields slope, n and intercept, k_1 at the particular applied load P that exists at $d = 10^{-3}$ m for any set of discrete data. The literature (Beigh et al., 1995) speaks that for any material with rising applied load H_v decreases when n<2 and increases when n>2. Both n and k_1 computed for the eutectic alloy Sn- Cd are recorded in Table 6.In fact, hardness is a measure of the resistance to localized plastic deformation, defined as the ratio W/A where W, the load in N and A, the area of indentation in m^2.This implies that a load applied to a specimen is partially affected, since a portion of it being utilized to overcome the Newtonian resultant pressure, W of the specimen itself. Consequently, the H_v data of the specimen necessarily bound to be analyzed in terms of the actual load (P- W) acting on the specimen. The inference from this concept to be drawn is that a load less than W will not result in plastic deformation. In view of the resultant pressure, Hays and Kendall (Hays & Kendall, 1973) modified Eq. (13) in a manner:

$$(P-W) = k_2 d^2 \tag{14}$$

where k_2 is a constant and n=2, in the Eq. (14) called the logarithmic index, since W allows the limiting case to prevail where hardness is not markedly dependent on the load and the index can be evaluated by subtracting Eq. (13) from Eq. (14), which results:

$$d^n = (k_2 / k_1) d^2 + W/ k_1 \tag{15}$$

again, the plot of Eq. (15) for d^n versus d^2 (Fig. 17) yields the slope k_2 / k_1 and the intercept W/k_1. From the known value of k_1, obtained by the plot (Fig. 16), k_2 and W computed thereby are recorded in Table 6 accomplishing the key data on microhardness. Furthermore, the plot (Fig.18) of log (P-W) versus log d of Eq. (14) yields the value of the logarithmic index, n < 2, virtually evincing the validity of the Newtonian resistant pressure concept of the eutectic alloy Sn- Cd.

The fracture toughness, K_c is a measure of resistance to brittle fracture when a crack is developed on the alloy specimen and follows the relation:

$$K_c = P/ \beta l^{3/2} \tag{16}$$

where l is the crack length measured from the center of the indentation mark to the crack tip. β is a numerical constant that depends on indenter's geometry, β = 7 for Vickers indenter. The satisfactory values of K_c can be obtained from Eq. (16) only if l /a > 3 or l /a < 3 respectively for median or radial crack system (Fig. 13 d) where a = d / 2. Table 7 records the parameter K_c and factor l /a at variable applied load P. The plot in (Fig. 19) shows the linear dependence of the crack length l with increasing applied load P. The brittleness is usually expressed as the brittleness index B_i and can be estimated from K_c values by the relationship (Lawn & Fuller, 1975):

$$B_i = H_v \ / \ K_c$$

$$(17)$$

B_i values calculated by Eq.(17) for the eutectic alloy Sn- Cd are presented in Table 7.The yields strength of a material can be estimated from the microhardness values using the relation (Beigh et al., 1995) valid for n<2:

$$\sigma_V = H_v \ /3$$

$$(18)$$

the yields strength of the eutectic alloy in the load range 10 X 10^{-2} – 100 X 10^{-2} N for constant indentation interval for 20 s is recorded in Table 7. All these parameters authenticate the plastic deformation of the eutectic alloy Sn- Cd and comprehensively accomplish the key data on hardness.

Figure 20 shows the XRD patterns for the eutectic alloy Pb-Bi containing a number of peaks and intensities of pure Pb and Bi atoms and expressing their immiscibility in the solidus eutectic, since the eutectic alloy does not indicate any peak of its own. The analysis implicitly, confirms the eutectic alloy to be a mechanical mixture of its constituent phases simulating weak interactions at their atomic levels. This implies that the eutectic composite is a terminal solidus solution.

CONCLUSIONS

Thermodynamic analysis of binary solidus-liquidus equilibrium data besides predicting the non-ideal nature of the systems, also offers an alternative thermal device for studying the phase equilibrium curve of binary systems, particularly capable of forming eutectic mixtures, in terms of excess functions G^E, S^E and H^E by variation in a mole fraction composition. The heterogeneous nucleation data verify the nucleation theory revealing the consistency in symmetry relations which predict that there is a definite relationship between the nucleation determined from the limit of undercooling and the liquidus structure of eutectic and non-eutectic melts associated with the entropy of fusion. The analysis of anisotropic growth kinetics emphasizes that the nucleation of eutectic phases from a binary melt follows the diffusion growth process. The rheological properties evince the essence of prominent molecular interactions in the binary eutectic melt and their emulation as a function of temperature. The excess functions and rheological properties are found complementary in predicting the non-ideal nature of binary eutectic systems and hence their liquidus structures. Microscopic studies confirm the f-f growth of the eutectic phases with $\alpha > 2$ and nf-nf growth of the eutectic phases with $\alpha < 2$, and also reveal their dependence on both temperature and modes of solidifications. The strength-growth relationship follows an identical form of the Weibull probability distribution curve that acquires two cut-off points corresponding to a lower strength limit in the slow and fast growth regions, and an upper strength limit in the moderate growth region. The moderate growth is an experimental evidence for being the most probable one in the domain of solidification modes in structuring the modal microstructure consisting of crystallites or lamellae reinforcing the matrix where there is a perfect crystallite-matrix bond. Of greater interest is the akin mass composition that predominates the domain of compositions in strength view point because of its superior microstructure. The chapter also discusses the micoardness of binary metallic eutectic system Sn-Cd and the experimental evidences confirm the plastic deformation of the system. The chapter work authenticates with experimentally investigated strength data that the physical properties, and in particular, the mechanical behavior of a material depend on the growth habits to produce the modal microstructure of the material. X-ray studies affirm the mechanical combination of the eutectic phases resulting in terminal solidus solution and hence the eutectic solidus structure.

ACKNOWLEDGEMENTS

The authors gladly express thanks and acknowledge the instinct cooperative spirit of Ms. Surby Gupta, Mr. Arun Kumar, Ms. Savita Gupta, Mr. Sahil Sharma (Computer Assistant), Mr. Rajesh and Mr. Asim Sharma in making the chapter work more effective as a teaching and learning, and research tool, and all those who have shared their input and contributions in compiling the chapter.

REFERENCES

1.   K. H. Ashbee, 1993 Fundamental Principles of Fiber Reinforced Composites, 2nd ed., Technomic Publishing Company, Lanchaster, PA (1993).

2.   C. Ascheron, C. Hasse, G. Kuhn, H. Neumann, 1989 Microhardness of Sn-doped InP. Crystal research and technology. 24 2 (February 1989), K33K35).

3.   S. Beigh, P. N. Kotru, B. N. Wanklyn, 1995 Indentation induced microhardness and fracture studies on (110) and (001) planes of flux-grown dysprosium orthoferrite single crystals. Materials chemistry and physics, 40 1995), 99104).

4.   H. E. Boyer, 1999 Hardness Testing, 2nd ed., ASM International Materials Park, OH (1999).

5.   C. A. Brookes, 1986 Inst. Phys. Ser, 75 Chapter 3.

6.   W. D. Callister, 2006 D.G. Rethwisch, 2008. Composites: Polymer-Matrix Composites. Fundamentals of Materials Science and Engineering, 3rd ed., Hoboken, New York (2008).

7.   W. D. Callister, Materials Science and Engineering. An introduction. 6 Ed., John Wiley and Sons Inc., Canada (2006).

8.   R. Caram, M.. Banan, W. R. Wilcox, 1991 Directional solidification of Pb-Sn eutectic with vibration. Journal of crystal growth, 114 July 1991), 249254).

9.   R. Caram, S.. Chandrashekher, W. R. Wilcox, 1990 Influence of convection on rod spacing of eutectics. Journal of crystal growth, 106 march1990), 294302).

10. T. H. Courtney, 2000 Mechanical Behavior of Materials, 2nd Ed., Mc Graw- Hill Higher Education, Burr Ridge, IL, (2000.)

11. W. W. Edward, 1930 International Critical Tables of Numerical Data, 7 McGraw-Hill,New York (1930).

12. H. W. Hayden, W. G. Moffatt, J. Wulff, 1965 The Structure and Properties of Materials, Vol. III, Mechanical Behaviour, John Wiley & Sons, New York (1965).

13. C.. Hays, E. G. Kendall, 1973 Analysis of knoop microhardness. Metallography. 6 August 1973), 275282).

14. D. Hull, T. W. Clyne, 2006 An introduction to composite materials, 2nd Ed., Cambridge University Press, Newyork, 2006.

15. D. Hunt, K. A. Jackson, 1966 Binary Eutectic Solidification. Trans. AMIE, 236 June 1966), 843852).

16. F. P. Kollman, F. P. Ed, Berlin. Kollman, 1975 Principles of Wood Science and Technology, Heildberg, New York (1975).10.1007/978-3-642-87931-9

17. B. Lawn, R., E. R. Fuller, 1975 Equilibrium penny- like cracks in indentation fracture. Journal of Material science. 10 (1975), 20162024).

18. D. R. Lide, 2009 CRC Handbook of Chemistry and Physics, A Ready Preference Book of Chemical and Physical Data:90thEd.,CRC Press,London 2009.

19. P. K. Mallick, 1993 Fiber-Reinforced Composites, Materials, Manufacturing, and Design, 2nd ed., Marcel Dekker, New York (1993).

20. D. E. Ovsienko, G. Alfinstev, A., T. Arizumi, 2 2nd (1980) Crystal Growth, properties and Application, 2 Springer-Verlag, Berlin Heidelberg New York 1980.]

21. R. M. Pillar, 1984 "Manufacturing Processes of Metals: The Processing and Properties of Metal Implants," Metal and Ceramic Biomaterials, Ducheyne, P & Hastings, G (Editors), Franklin Book Company, Elkins Park, PA (1984).

22. B. L. Sharma, 2003 Structural models of faceted-faceted eutectic system vanillin-acenaphthene Mater. Chem. Phys. 78 (2003), 691701).

23. B. L. Sharma, R. Jamwal, R. Kant, 2004a Thermodynamic and lamella models relationship for the eutectic system benzoic acid-cinnamic acid Cryst. Res. Technol. 39, 5 (2004), 454464).

24. B. L. Sharma, S. Tandon, R. Kant, R. Sharma, 2004b Quantitative essence of molecular interactions in binary organic eutectic melt systems Thermochimica Acta 421 (2004), 161169).

25. B. L. Sharma, S. Gupta, S. Tandon, R. Kant, 2008 Physico-mechanical properties of naphthalene- acenaphthene eutectic system by different modes of solidification. Mater.Chem.,Phys.111 (2008) 423430)

26. B. L. Sharma, S. Tandon, S. Gupta, 2009 Characteristics of the binary faceted eutectic : benzoic acid-salicylic acid system, Cryst. Res. Technol. 44, 3 (2009). 258268)

27. W. A. Woishnis, 1993 Engineering Plastics and Composites, 2nd ed., ASM International Materials Park, OH (1993).

Chapter 8

IN-SITU DEPOSITION OF METAL MATRIX COMPOSITE IN FE-TI-C SYSTEM USING LASER CLADDING PROCESS

Ali Emamian, Stephen F. Corbin and Amir Khajepour
University of Waterloo, Canada

INTRODUCTION

Composite materials result from the combination of two or more dissimilar materials with different physical and mechanical properties. The final product has superior properties compared to the individual components. Particles or reinforcements in different geometries, including particulates, fibres and whiskers are used in various types of matrices such as polymers, ceramics or metals. The toughness and strength of composites are thus functions of the matrix and reinforcements properties and phase morphology.

Most composites are made from a ductile matrix in which hard particles are distributed. The matrix of a composite can be a polymer, ceramic, or metal. Metal matrix composites (MMCs) are a type of composite in which ceramics, such as TiC, WC and TiB_2 with a high melting point and high hardness, are distributed in a metal matrix like Fe, Co or Ni. Fe and its alloys have attracted substantial attention of late due to their advantages over other alloys. These advantages include availability, low cost, and chemical compatibility with a wide range of steels. Al_2O_3, ZrO_2, TiN, TiC and VC are examples of hard particles used as reinforcements in metal matrices. Among these particles, TiC has sparked considerable interest because of its good wettability with Fe, high hardness and low density compared to the other components. In metal matrix composites, hard particles distributed in the matrix cause an increase in strength, stiffness, wear resistance and decreased density (Cui et al., 2007; Emamian et al., 2010c; Pu, 2008).

Different methods are used to form Fe-TiC composites. For example, Das et al. (2002) reviewed various synthesis routes of TiC-reinforced Fe-based composites such as powder metallurgy, conventional melting, casting, combustion synthesis and laser surface melting. Additionally, the authors formed TiC by a combustion synthesis method known as aluminothermy. Similarly, Jiang et al. (2000, 2007) fabricated a metal matrix composite by a casting process. Using WC and TiC as reinforcing particles, they determined that these particles are evenly distributed in the metal matrix. Unfortunately, these methods are not well suited to produce an MMC coating.

Laser cladding is an effective method which is extensively used in surface modifications of metallic materials. The highly-focused generated heat in this process results in a minimal heat-affected zone. Therefore, the process has a minimum impact on a substrate's mechanical properties. In conventional laser cladding, the majority of reinforcing phases are directly added into the coating materials (i.e., an ex-situ process). Ariely et al. (1991) carried out laser alloying of AISI 1045 steel with TiC powder fed by the dynamic blowing method. They studied the effect of changing the laser power, scan speed and feed rate on the process. While optimum parameters significantly increase the surface hardness, some solubility of TiC in the molten Fe produces a small fraction of TiC dendrites upon re-solidification of the coating. Tassin et al. (1995) used a laser process to enhance the surface hardness of AISI 316L by adding TiC particles. As well, they added chromium carbide (Cr_2C_3) instead of TiC to augment the hardness by as much as 450 to 900 HVN.

In the ex-situ method, chemical incompatibility between the reinforcement and matrix can occur, leading to poor interfacial bonding. Different thermal properties between the composite phases (e.g., coefficient of thermal expansion, CTE) increase the risk of cracks at the poorly bonded ceramic/ matrix interface in ex-situ synthesis. In contrast, in-situ laser cladding can eliminate the interface problem. The in-situ process is a method where the hard ceramic phase is formed during coating by the reaction between powder ingredients. This creates a thermodynamically stable ceramic and matrix phase with sufficient strength to transfer stresses. Therefore, the probability of crack formation and failure at the composite interface decreases. Moreover, the high solidification rate in laser cladding opens the option of controlling particle size during the evolution of reinforcement particles.

The degree of uniformity of the particle distribution throughout the matrix directly affects the degree of uniformity of the composite properties. It is therefore important to control particle distribution during a in-situ laser cladding process. The distribution of phases and the overall clad microstructure is sensitive to the relatively large number of laser cladding operating parameters. These parameters, as well as the physical phenomena which govern the in-situ process and determine the final quality of the fabricated parts, make this layer-based technique a complex process.

Du et al. (2007a, 2007b) enhanced the wear resistance and hardness of low carbon steel through the formation of TiC-VC particles in an iron matrix by laser cladding. They also achieved a similar result for $TiC-TiB_2$. The increase in hardness was due to carbides being distributed uniformly in the metal matrix. Wu et al (2001) increased the surface hardness with TiC_p/Ni using laser cladding.

Wang et al. (2008, 2009) increased the wear resistance of AISI 1045 steel by the in-situ synthesis of a composite coating by preplacing FeCrBSi alloy and graphite powders on the substrate surface prior to the laser surface treatment. They proposed that increasing the wear resistance might be due to the formation of a variety of in-situ carbides including TiC particles. The in-situ process was also selected by Yan et al. (1996), who formed TiC in a nickel-based alloy coating on mild steel. During the cladding process, other types of particles such as $Cr_{23}C_6$, Ni_5Si_2 and Cr_2B were formed and uniformly increased the hardness.

From the above literature review, it is clear that the in-situ formation of TiC particle metal matrix composites through laser cladding is a feasible process. Furthermore, it is widely reported that the hardness and wear resistance of the substrate can be significantly enhanced by the formation of a TiC-based composite coating. It is also clear that the TiC morphology can vary throughout the clad thickness, but no detailed or satisfactory explanation on why this happens, or what the mechanism of TiC formation is during laser cladding, has been published. This is partly due to the lack of knowledge and understanding about the relationship between clad microstructure and laser processing conditions. In addition, most researchers have focused on the use of rather complex, multi-component metal matrix powders containing various combinations of Ni, Fe, Co, Cr, B or Si. This can produce a variety of carbides other than TiC and creates complex solidification behaviour during cooling.

This chapter discusses the development of an in-situ laser cladding technique to deposit a TiC-based MMC coating on a steel substrate with no cracks and excellent bonding to the substrate and a high hardness. The approach involves feeding graphite, titanium and iron powders into a melt pool to form an in-situ clad consisting of TiC in a Fe-rich metal matrix. To study the effects of laser parameters on the quality of the clad, they are combined in two general parameters (effective energy and powder deposition density) to gain a better understanding of their roles. These parameters can help to predict the clad quality prior to conducting experiments (Emamian et al., 2009).

In order to study the effect of chemical compositions on TiC morphology, micro hardness and wear resistance of the clad, diverse chemical compositions resulting from different C:Ti ratios (i.e., 45:55 and 55:45at%) and Fe content (i.e., 70, 60, 50 and 10 wt%) were selected for the powder composition. Additionally, several laser parameters were selected to study their effects on the clad microstructure.

Phase Diagram

In this research, Fe, Ti and C are the primary components. The binary and ternary diagrams help to find the phases which are expected to appear during the process. Although the laser cladding process is a non-equilibrium process, phase diagrams are useful guides to interpret and estimate the result-ant phase formation.

TI-C Phase Diagram

The selection of the atomic ratio forming a TiC phase is important in the formation of the TiC particles (see Fig. 1).

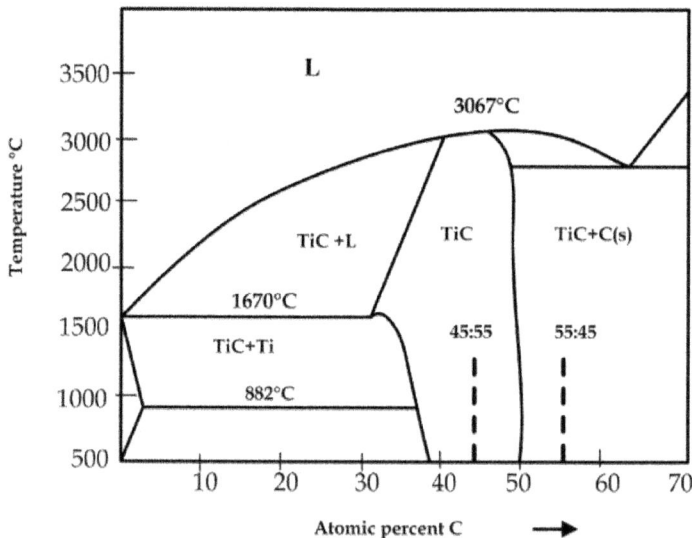

Figure 1: Titanium-Graphite phase diagram (Hirokai, 2000).

It can be seen that TiC is stable over a wide range of composition as a intermediate phase. To avoid the formation of secondary phases such as Ti (α), Ti (ß) or C, a composition close to Ti-45%at C should be selected. For the composition study section C:Ti, the composition was changed to 55:45. Obviously, in the non-equilibrium condition phase formation can deviate from the predictions of the equilibrium phase diagram. Nonetheless, these diagrams are very useful in predicting, at least qualitatively, phase formation under non-equilibrium states.

FE-TI Phase Diagram

Since Fe and Ti are in contact during the melting of non-equilibrium solidification, the phase diagram below should be considered. Titanium is a ferrite stabilizer and austenite is stable in a narrow zone.

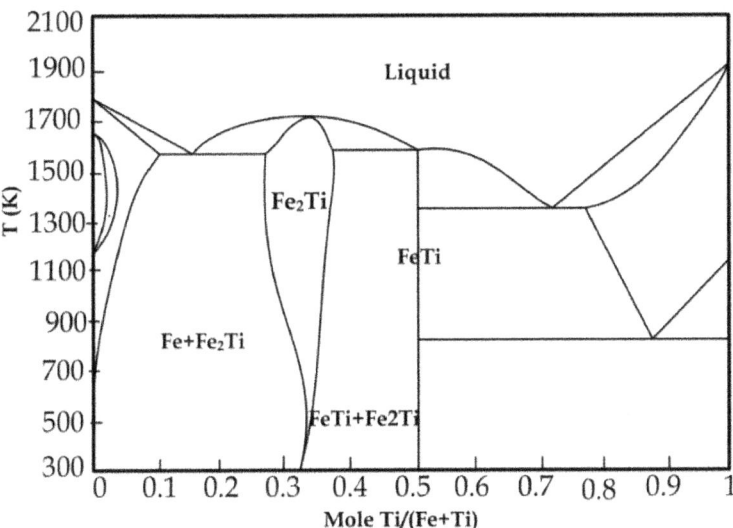

Figure 2: Iron-Titanium Phase diagram (Hirokai, 2000).

According to Fig. 2, there are two types of intermetallics: Fe_2Ti and $FeTi$. Neither shows acceptable mechanical properties and hence are undesirable phases. The phase diagram of Fe-Ti shows that, by moving to the right of the phase diagram, the likelihood of both intermetallic phases being present increases. This is particularly so between the 0.3-0.5 mole fractions of titanium, where both intermetallics are in equilibrium. The goal is to form a metal matrix composite and decrease the chance of an intermetallic formation.

Thermodynamics of Formation

Phase transformation calculations indicate that it is highly probable to have TiC particles instead of any other carbide such as Fe_3C in an in-situ TiC-Fe cladding.Fe_3C is one of several compositions which can be formed in the melt pool. The following thermodynamic equations provide information about the reactions (Dubourg & St-Georges, 2006):

$$[Ti] + [C] = <TiC> \quad \Delta G = -139,0461.31 + 98.96 \text{ T}$$
$$j / mole \tag{1}$$

$$3Fe_{(l)} + <C> = <Fe_3C>, \quad \Delta G = -53416 + 55.68\,T$$
$$j / mole \tag{2}$$

The equations prove that the Gibbs free energy of TiC formation is more negative than that of Fe_3C. Therefore, the probability of Fe_3C formation is lower than TiC. Although the equations are useful in equilibrium conditions, they are still useful to predict the formed phases. Moreover, it can be seen in the next chapter that there are no results to confirm the presence of Fe_3C.

EXPERIMENTAL SET-UP

In the experiments, AISI 1030 medium carbon steel is chosen as the substrate, with 0.28-0.34°C, 0.6-0.9 Mn, P less than 0.04, and S less than 0.05 in wt%. The substrate dimensions on which the claddings are deposited are 100×30×6.35 mm. Prior to laser cladding, the substrates are shot-blasted and rinsed with ethanol, followed by acetone.

Mixtures of Ti, graphite, and Fe powder are fed into a melt pool formed by the laser using a Sulzer Metco (TWIN 10-C) powder feeder. The ratio of C: Ti is 45:55 at %C and Ti-55 at %C. A three-digit code is considered for powder compositions. The first digit stands for Fe percentages (7 for 70) and the next two digits are the C:Ti ratios in atomic percentages (i.e., 45 or 55%). In the first two sections presented below, the powder composition was set at Ti-45% at C (745). In the composition study section, the remainder of the samples are studied (i.e. 755, 655, 555, 155). Table 1 presents the chemical compositions of powders which were studied in this chapter.

Mixtures of the powders are pre-cursors, with 99.5% purity; a 40 μm maximum powder size is further mixed with 98% purity Fe powder. To obtain a homogeneous distribution, all powders are blended for four hours at 300 rpm in a quarter-filled jar of 165×60 mm with a milling media of glassy balls where the ball-to-powder-size ratio is 20:1.

Table 1: Chemical composition of investigated powders

Sample group code	Fewt %	Ti wt%	C wt%	Fe at%	Ti at%	C at%
Group 745	70	25	5	57	24	19
Group 755	70	23	7	55	20	25
Group 655	60	31	9	44	26	30
Group 555	50	38	12	33	30	37
Group 155	10	69	21	5	43	52

An IPG fiber laser model YLR-1000-IC operating in continuous mode with a maximum power of 1 kW is used to produce a series of single-clad tracks. The diameter of the laser beam spot size on the workpiece (WP) surface is fixed at 2.5 mm. The laser machine is integrated with a five-axis CNC vertical machining centre to control the velocity of the WP. To protect the melt pool from oxidation, Argon shielding gas is supplied through a nozzle at 10 Lmin^{-1}.

The specimens are sectioned for micro-structural examination in a longitudinal direction. The samples are prepared using SiC grit paper with grit sizes ranging from 240 to 2400 and polished with diamonds from 6 microns to 1 micron. Afterwards, the samples are placed in an ultrasonic bath to remove any surface contaminants and then rinsed with alcohol and air dried.

A LEO SEM with 20keV is used to examine the microstructure and morphology of the phases formed during cladding. SEM is selected to determine the quality of the clad deposit (i.e., bonding to the substrate) and to determine the in-situ TiC particle morphology. The working distance is 9-14 mm and the vacuum system is set at 1.53e-0.005 mBar. Different magnifications are used to capture the microstructure. The mounting material used was conductive in order to prevent charging during sample observation.

The micro-hardness at different depths from the substrate-clad interface is measured using a Vickers micro-hardness tester, and the average of three hardness measurements is reported.

A low magnification EDS analysis of the whole longitudinal section of the clads helps to measure the Fe percentage to calculate the dilution. Further, an XRD analysis is performed with a Rigaku AFC-8 diffractometer.

Clad heights were measured by a caliper vernier. Specifically, measurements were made at the middle of the clad which normally were uniform in height. Clad height was also measured using scale bars in low magnification SEM images.

LASER PROCESSING CONDITIONS

The laser process parameters to determine the optimum laser cladding conditions are given in Table 2. Initial experiments were held at a constant powder feed rate of 8 g/minwhile systematically increasing the work piece (WP) scan speed from 2,4 to 6 mm/s. Each scan speed was explored at laser powers from 250 to 650 W (i.e., samples 1 to 9). Visual inspection of the formed clad tracks revealed that, for the power setting, a scan speed of 6 mm/s was required to produce a clad deposit.

In the second series of experiments (samples 10 to 13), the scan speed was increased from 6 to 16 mm/s while fixing the laser power and powder feed rate at 650 W and 8 g/min., respectively. Again, a clad deposit was produced in each case (except at the highest scan speed) but no bonding with the substrate occurred.

In an effort to create a clad/substrate bond, the laser power was further increased to 700 and 800 W while fixing the scan speed at 6 mm/sand using powder feed rates of 4 and 8 g/min (samples 14 to 17). Partial bonding of the clad to the substrate occurred in these cases. For example, at the higher powder feed rate, the clad adhered to the substrate immediately after cladding but detached during complete cooling. For the 4 g/min feed rate, clads remained attached to the substrate but were easily removed upon application of a small force by hand. Lowering the WP scan speed to 4, 3 or 2 mm/s(i.e., samples 18-20) did not rectify this situation.

In the final series of experiments, the laser power was increased to 900 W while the powder feed rate was at 4 or 8 g/min and the scan speed was at 6 or 8 mm/s. A laser power of 1000 W was also used with a scan and powder feed rate of 4 mm/s and 4 g/min, respectively. Visual examination of these clad tracks revealed that in all cases a clad deposit was produced and was well bonded to the substrate.

Sem Results

SEM is selected to determine the quality of the clad deposit and bonding with the substrate and to find out if in-situ TiC particles are formed. For this investigation, samples which indicated a good clad deposit but no bonding with the substrate (i.e., samples 3 and 9) and a successful clad deposit (i.e., samples 23 and 24) were chosen.

However, none of these processing conditions were able to create a bond between the clad deposit and substrate.

Fig. 3 illustrates an SEM back scatter image of a typical region from the un-bonded clad deposit sample 9. It is clear that the clad microstructure consists of a matrix in which a relatively high volume fraction of uniformly dispersed particles are present. The size of these particles ranges from approximately 1 to 7 m.

Table 2: Results and Observation

No	Power (W)	Scan Speed(mm/s)	Feed Rate (g/min)	Effective Energy (J/mm2)	Powder Deposition Density (g/mm2)	Observation
1	250	2	8	50	0.027	No Bond-No Clad
2	250	4	8	25	0.013	No Bond-No Clad
3	250	6	8	17	0.009	No Bonding-Formed clad
4	400	2	8	80	0.027	No Bond-No Clad
5	400	4	8	40	0.013	No Bond-No Clad
6	400	6	8	27	0.009	No bonding-Formed clad
7	650	2	8	130	0.027	No Bond-No Clad
8	650	4	8	65	0.013	No Bond-No Clad
9	650	6	8	43	0.009	No Bonding-Formed Clad
10	650	8	8	32	0.007	No Bonding-Formed Clad
11	650	10	8	26	0.005	No Bonding-Formed Clad
12	650	12	8	22	0.004	No Bonding-Formed Clad
13	650	16	8	16	0.003	No Bond-No Clad
14	700	6	8	47	0.009	Partial bonding with substrate melting

15	700	6	4	47	0.004	Weak bond-Removable Clad
16	800	6	8	53	0.009	Partial bonding with sub-strate
17	800	6	4	63	0.004	Weak bond-Removable Clad
18	800	2	8	160	0.027	No bond-ing-Formed Clad
19	800	3	8	106	0.018	No bond-ing-Formed Clad
20	800	4	4	80	0.007	Partial bonding with sub-strate
21	900	6	8	60	0.009	Successful Clad
22	900	8	8	45	0.007	Successful Clad
23	900	6	4	60	0.004	Successful Clad
24	900	8	4	45	0.003	Successful Clad
25	900	4	4	90	0.007	Successful Clad
26	1000	4	4	100	0.007	Successful Clad
27	1000	4	4	100	0.007	Successful Clad

Fig. 3 also illustrates a higher magnification image of a eutectic structure in grain boundaries of the matrix of sample 9 (i.e. region 2). Table 3 reports a typical EDS analysis performed on the dark grey dispersed particles and regions 1 and 2 of the matrix. The high Ti content of the particles is consistent with TiC particles. The majority of the matrix is a very Fe-rich phase with slightly elevated Ti concentrations in region 2. The very small black particles in Fig. 3 were too small to obtain reliable EDS analysis.

Carbon was not included in the analysis due to inaccuracies in its inclusion in a quantitative analysis.

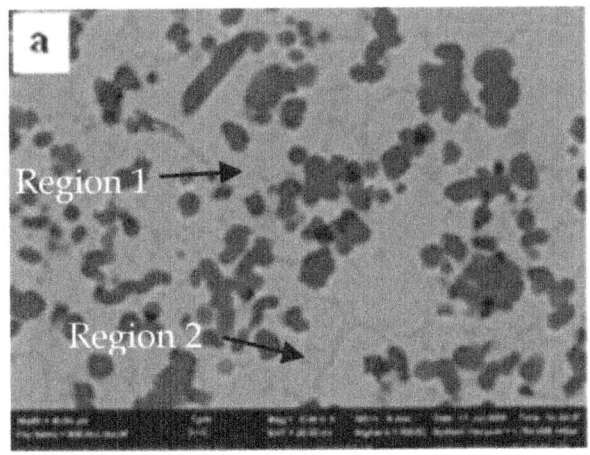

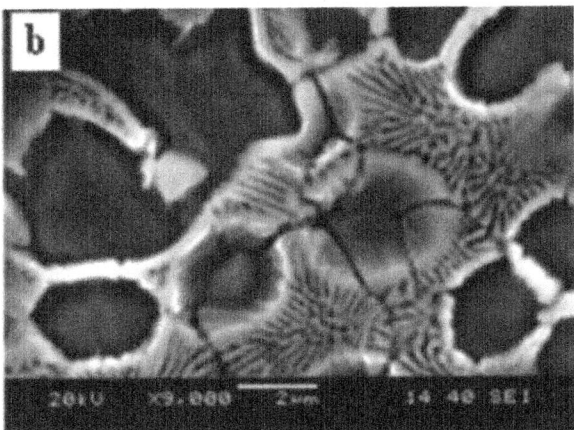

Figure 3: BSE image of sample 9 a) distributed black particles in individual and longitudinal shape in matrix of un-bonded clad b) High magnification of Region 2 Eutectic structure of Fe-Fe$_2$Ti.

Table 3: EDS analysis of phases in sample 9

Region	Ti conc.(wt%)	Fe conc.(wt%)
Dark grey particles	95.2	4.8
Region 1	8.7	91.3
Region 2	16.5	83.5

Fig. 4 shows a high magnification image of the two phase microstructure of sample 23. The microstructure consists of a high volume fraction of fine (i.e., < 2 m) particles dispersed in a matrix. As with Table 3, EDS analysis confirmed that the dark grey particles are Ti-rich, whereas the matrix is Fe-rich.

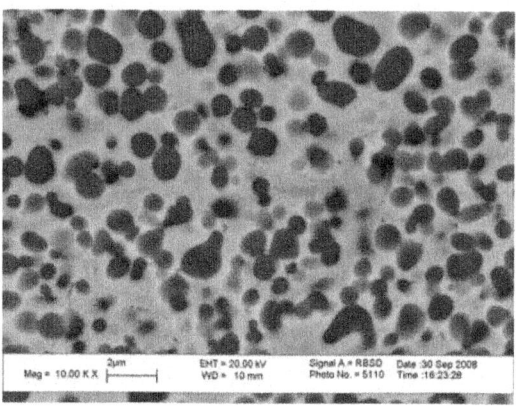

Figure 4: BSE image-distributed black particles in matrix of a one layer clad (sample 23).

Fig. 5 shows a longitudinal section of sample 24 which confirms a uniform distribution of Ti-rich particles in an iron-rich matrix. However, in some areas, such as those labeled region A, larger black particles exist. Qualitative EDS analysis of the particles, which included carbon in the analysis, indicated a 95 wt% carbon composition with the balance as Ti and Fe. Note that the black particle in the centre of the image is surrounded by the Ti-rich grey phase. An EDS analysis of this region reveals an approximately 95 wt% Ti: 5 wt% Fe composition.

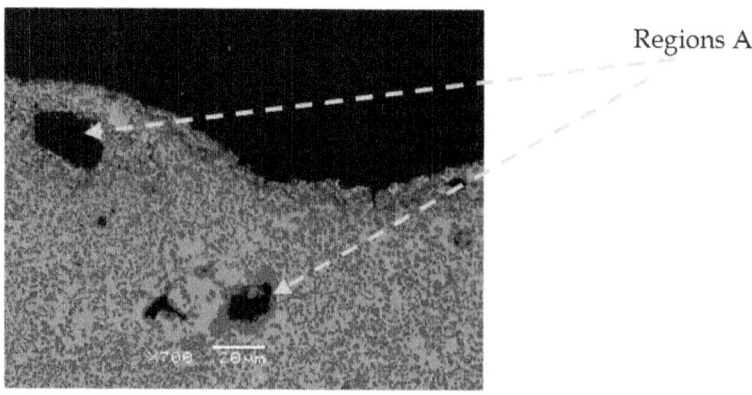

Regions A

Figure 5: Sample 24 TiC distributions.

XRD Results

XRD analysis was performed on samples 9 (un-bonded clad) and 23 (one layer clad). The XRD spectrums for these samples are shown in Fig. 6.

Fig. 6 depicts the XRD result for the un-bonded clad. According to this graph, TiC, Fe_2Ti and iron-titanium solid solution exist in the clad. Since there is no bonding with the substrate, dilution does not contribute to the results.

The X-ray diffraction pattern for a one layer clad is plotted in Fig. 6 and confirms the presence of the iron-titanium solid solution and TiC in the clad zone. It is worth noting that the intermetallic phase (Fe_2Ti) did not appear in the diffraction peaks of the clad coating, in contrast to the result for the un-bonded case.

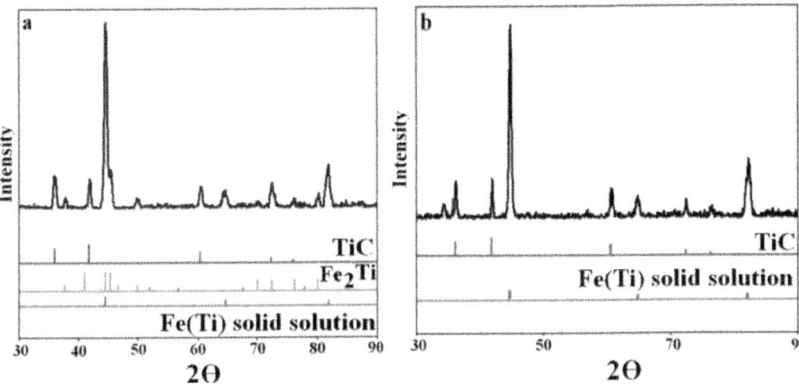

Figure 6: XRD result of a) un-bonded clad (sample 9) and b) one layer clad (sample 23).

Further Laser Process/Clad Development and Analysis

From the data in Table 2, it is clear that laser power scan speed and powder feed rate all play a vital role in clad formation, bonding, and clad quality. From the table, we can see that increasing the power increases the probability of forming a high quality clad. However, it is not the only effective parameter. In any period of time, the laser energy must provide sufficient heat to melt a given volume of the substrate as well as the incoming powder stream. This will be determined not only by the laser power but also the work piece (WP) scan speed, and powder feed rate. Two combined parameters (i.e., the effective energy per unit area (*Eeff*) and the powder deposition density (*PDD*)) are needed in order to analyze the influence of these three parameters.

Effective energy is defined as the parameter which provides a measure of the delivered energy to the process by the laser. This energy is principally responsible for melting the substrate surface and powder and can be defined by computing (Emamian et al., 2009, 2010a).

Energy per unit area (J/ mm²) = P/(V.D)

(3)

where P is the laser power, V is the scan speed of WP, and D is the laser spot size in *mm*. Therefore the units of effective energy are W ($mm^{-2}s^{-1}$) or Jmm^{-2}.

The powder deposition density (*PDD*) is introduced as (Emamian et al., 2010a, 2010b):

Powder deposition density (g/ mm²)= R/(V.D)

(4)

where V and D have the previous meaning, R is the powder feed rate in g/min, such that the powder deposition density (*PDD*) has units of g/mm².Table 2 reports these calculated parameters for each of the laser conditions. Fig. 7 plots the data where *PDD* is the x-axis and (*Eeff*) is the y-axis.

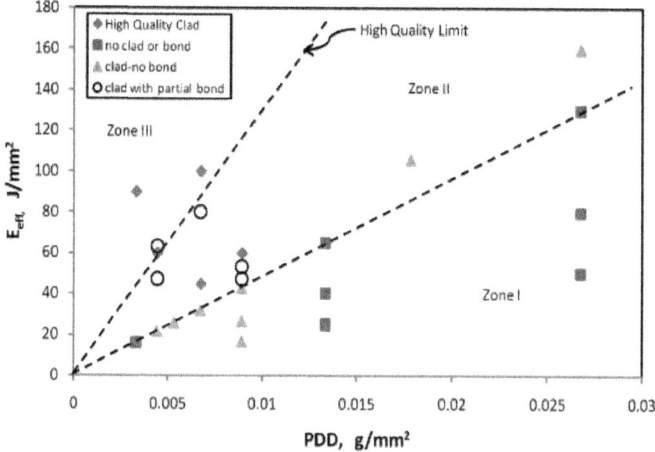

Figure 7: Effective energy versus powder deposition density.

As seen in Fig. 7, low *Eeff* and high *PDD* values leads to no clad deposit formation or bonding with the substrate. (i.e., Zone I). While increasing the *Eeff* and reducing *PDD* leads to the formation of a clad deposit, it is not bonded to the substrate (i.e., Zone II). Finally, high effective energy along with lower powder deposition density leads to the formation of a dense clad deposit which is well bonded to the substrate (i.e., Zone III). Therefore, to achieve an acceptable clad quality, effective energy and powder deposition density should be adjusted for Zone III. This graph helps us to predict the proper parameters to create an acceptable clad prior to conducting any experiment.

To further explore the predictability of this figure, a range of clad conditions which lie close to the boundary line between Zones III and II were chosen and the clad experiments completed. These conditions are discussed in the next section. All of the conditions produced a clad deposit that could not be removed from the substrate, indicating good bonding. Metallographic examination of these samples was completed to further analyze clad quality and microstructure.

TiC Formation and Morphology

In this section, the TiC formation mechanism is explained based on observations and phase diagrams. This explanation can be followed in two steps–namely, melting and solidification during laser cladding.

In order to explain the TiC morphology formed during laser cladding, the melting stage of the process must first be considered. When the powder mixture enters the laser beam, the powder will rapidly heat up (Fig. 8a). The melting points of pure Fe, Ti and C (or graphite) are 1538, 1668 and 3400°C, respectively. Given the relative melting temperatures of the pure powders, it is reasonable to assume that Fe melts first, creating a liquid phase that spreads and surrounds the Ti and graphite powders (e.g.,Fig. 8b). The interaction of the elements of the powder stream will cause the formation of a ternary mixture. Fig. 9 presents the liquidus surface projections for the entire ternary C-Fe-Ti composition. The actual ternary composition of the powder mixture deposited on the 1030 steel substrate is indicated by the black dot (745 PC) in Fig. 9. From these indicators, it is clear that both Ti and C have significant solubilities in the Fe-rich liquid phase and that this solubility increases with an increase in temperature. It is also clear from Fig. 9 that the Fe-rich liquid phase will be in equilibrium with solid TiC. Therefore, alloying Fe with Ti and C actually creates a ternary liquid through the gradual dissolution of the Ti and graphite powders.

In addition, a TiC layer will form between this ternary liquid and the original graphite particles in order to establish the required phase equilibrium. As the clad temperature increases, dissolution of the solid phases will continue as the solubility of Ti and C in the liquid increases. Also the growth of the TiC phase toward the graphite core will increase (e.g., Fig. 8-c). Evidence for the microstructure depicted in Fig. 8-c is presented in Fig. 5, which illustrates graphite particles surrounded by a TiC layer.

The next stage in clad microstructural development largely depends on the peak temperature induced by the laser. If the peak melt pool temperature exceeds the liquidus temperature of the ternary composition, all of the TiC/C particles will dissolve into the melt, creating a completely liquid melt pool. If no dilution by the substrate is assumed, then it can be stated that the liquidus temperature of the clad melt pool for composition of 745 is 2345°C. If the peak melt temperature is less than the liquidus temperature of the melt pool clad (e.g., in the presence of Fe substrate dissolution), then a semi-solid mixture will persist where the solid graphite particle core will eventually convert through the further development of the diffusion couple to TiC.

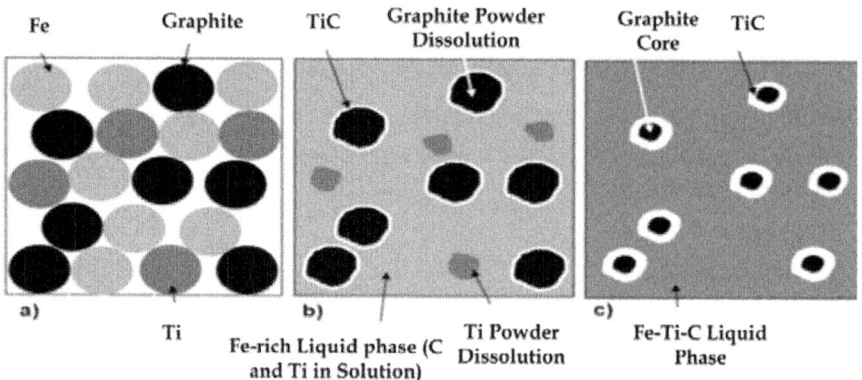

Figure 8: TiC formation mechanism.

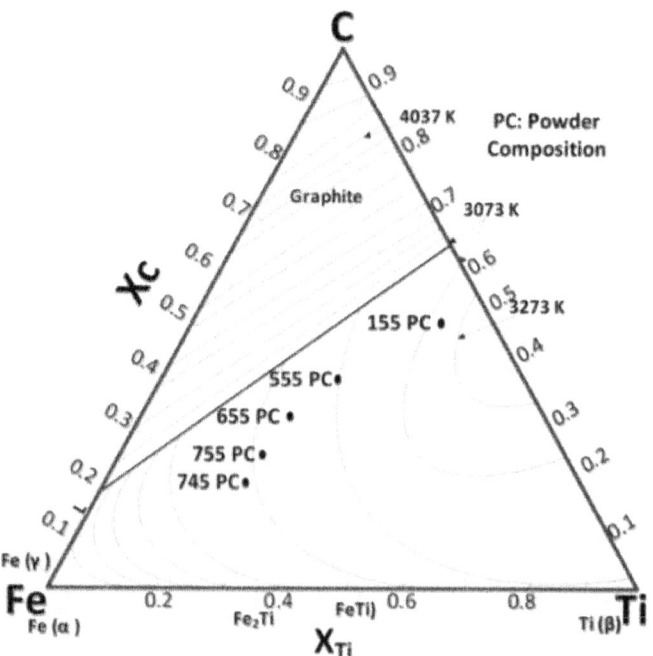

Figure 9: Liquid projection.

Fig.10 confirms that there is a large phase field where both liquid and solid TiC will exist in a semi-solid state. It also confirms that Fe-rich mixtures (e.g., the left-hand side of the diagram) have lower liquidus temperatures. Dilution with the substrate increases the Fe content in the clad, thereby decreasing the liquidus temperature. Dilution thus plays a crucial role in establishing the clad microstructure and TiC morphology.

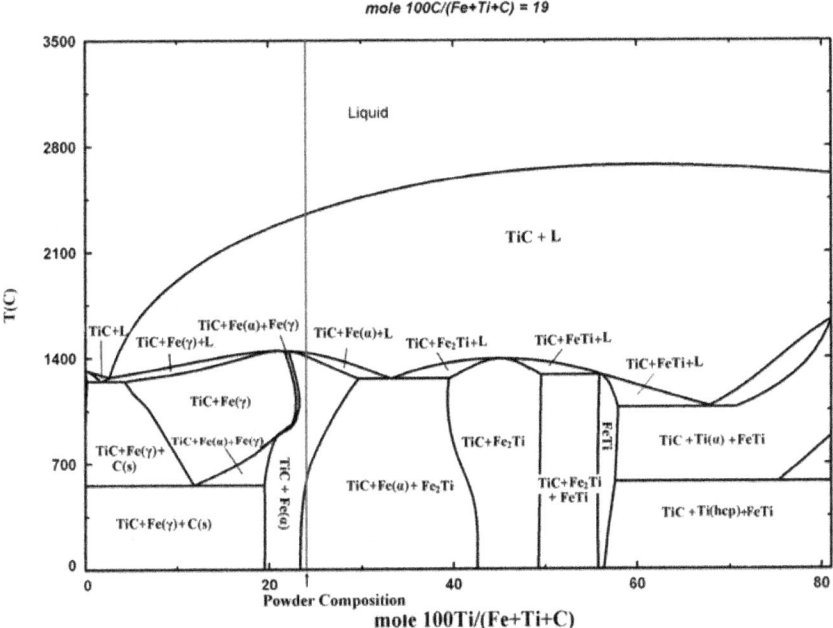

Figure 10: Cross section of Fe-Ti-C ternary diagram in 19%at C.

EFFECT OF COMBINED PARAMETERS ON TIC MOR-PHOLOGY AND CLAD MICROSTRUCTURE

In this section, the effect of laser parameters on TiC morphology is studied. Results show that laser parameters play a crucial role in the morphology of TiC. Dendritic or spherical TiC particles with different distributions are observed, depending on the applied laser parameters. Two combined parameters, effective energy and powder deposition density, are used in order to study the effect of laser process parameters on TiC morphology.

A series of experiments are conducted in constant laser power and scan speed, constant effective energy and constant powder deposition density in order to study the TiC morphologies. Laser parameters are selected in order to be in a high quality zone as well as to result in particular values for combined parameters (namely, effective energy and powder deposition density). These series of samples help to confirm the high quality zone (e.g., zone III) found in the previous section. Fig.11 depicts the process map for the experiments.

Samples with powder deposition density (*PDD*) greater than or equal to 0.010 g/mm² were selected for detailed analysis. Low clad height was the main reason for the lack of ability to analyze the samples with low *PDD* (less than 0.010 g/mm²).

Constant Laser Power with Different Scan Speeds (Type A Experiments)

Two groups of samples were investigated in this section: samples A1, A2, and A3, with 961 W laser powers at a scan speed of 120, 210, and 330 mm/min (Fig.11). The second group has 907 W laser power explored with scan speeds similar to the first group.

As observed in Fig.12, sample A1 shows dendrites of TiC particles in the clad. Table 4 shows that sample A1 has the maximum value for *Eeff* and *PDD* among the A series of experiment. As well, it exhibits the lowest value of dilution and highest clad height amongst the three samples. In samples A2 and A3, there is no evidence of dendrites, and the TiC particles are very fine. TiC particle size in sample A2 is in the range of 1-4 μm, while sample A3 has a smaller size range.

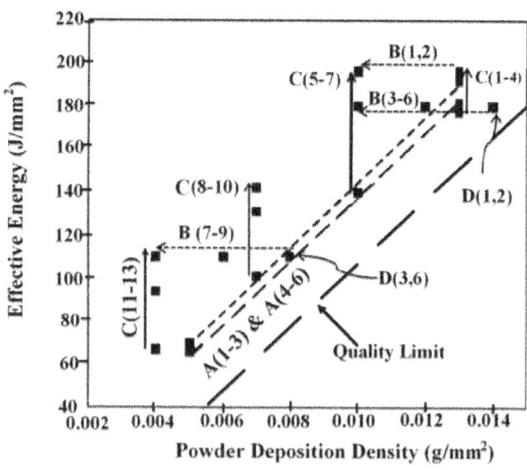

Figure 11: Map study for groups A to D.

The same trends were observed in samples A4, A5 and A6. In both groups of samples, it is noted that, by increasing the scan speed, dilution is increased, clad height is decreased, and the TiC microstructure is finer. This can be attributed to the relationship between scan speed and *PDD*. Specifically, as the scan speed increases, the *PDD* decreases. Since the laser power is constant, a larger portion of energy can be assigned for melting the substrate by decreasing the *PDD,* which then increases dilution.

Sample Category	Sample NO.	Laser Power (W)	Scan speed (mm/min)	Powder feed rate (g/min)	Effective Energy (J/mm2)	Powder deposition density(g/mm2)	Hardness (HVN)	Dilution %	Clad Height (mm)
A	1	961	120	4	192	0.013	575	72.8	1.40
	2	961	209.7	4	110	0.008	317	89.0	0.40
	3	961	330	4	70	0.005	750	94.9	0.14
	4	907	120	4	181	0.013	600	69.6	1.00
	5	907	210	4	104	0.008	400	88.0	0.10
	6	907	330	4	66	0.005	570	93.0	0.07
B	1	982	120	4	196	0.013	350	73.0	1.10
	2	982	120	3	196	0.010	650	77.0	0.60
	3	884	117	4	180	0.014	650	69.1	1.60
	4	907	120	4	181	0.013	600	69.6	1.00
	5	758.2	101.1	3	180	0.012	900	64.0	0.60
	6	589.7	78.6	2	180	0.010	2400	66.7	0.50
	7	961	209.7	4	110	0.008	317	89	0.4
	8	926.7	202.2	3	110	0.006	550	95	0.2
	9	864.9	188.7	2	110	0.004	600	---	---

C	1	884	120	4	177	0.013	650	70.3	1.60
	2	907	120	4	181	0.013	600	69.6	1.00
	3	961	120	4	192	0.013	575	72.8	1.40
	4	982	120	4	196	0.013	350	73.0	1.10
	5	700	120	3	140	0.010	850	70.0	0.90
	6	589.7	78.6	2	180	0.010	2400	66.7	0.50
	7	982	120	3	196	0.010	650	77.0	0.60
	8	506	120	2	101	0.007	644	79.0	0.07
	9	982	180	3	131	0.007	593	63.0	0.04
	10	708	120	2	142	0.007	584	75.5	0.1
	11	506	180	2	67	0.004	240	79.0	Low
	12	708	180	2	94	0.004	1000	80.0	0.14
	13	864.9	188.7	2	110	0.004	600	--	---
D	1	884	117	4	180	0.014	650	69.1	1.60
	2	663.4	88.5	3	180	0.014	1800	65.0	0.70
	3	961	209.7	4	110	0.008	317	89.0	0.4
	4	720.7	157.3	3	110	0.008	932	63.0	0.25
	5	480.5	104.8	2	110	0.008	450	88.8	0.10
	6	961	210	4	110	0.008	317	88.0	0.40

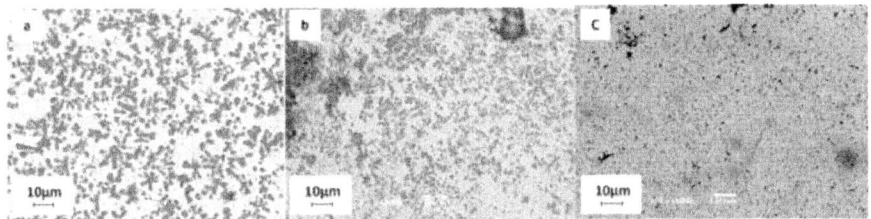

Figure 12: Sample A1 (dendrites of TiC), b) sample A2, c) sample A3.

Constant Effective Energy with Variable Powder Deposition Density (Type B Experiments)

The results of samples (e.g., sample B1 and B2 with 196 J/mm² and sample B3, B4, B5 and B6 with 180 J/mm²) are presented in this section. Fig.13 a-b illustrate the TiC morphology in samples B1 and B2, respectively. These samples have identical *Eeff* (i.e., 196), laser power and scan speeds, but their respective *PDD* values are 0.013 and 0.010, which result from different powder feed rates (4 and 3 g/min). This shows that sample B1 has a dendritic structure in almost the entirety of the clad, whereas sample B2 has very fine TiC particles distributed in the clad, as illustrated at Fig.13-b.

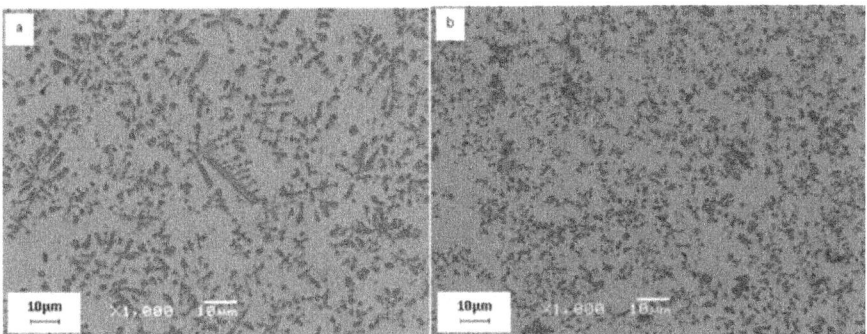

Figure 13: BSE images for a) sample B1 and b) sample B2.

Fig.14 a-b show the clad in samples B3 and B4. These samples have different *PDD*, 0.014 and 0.013. Sample B3 with *PDD* of 0.014 shows the dendritic structure of the TiC in the clad. At the bottom of the clad, very fine TiC particles can be seen approximately 100 μm from the clad/substrate interface. Fig.14-b also shows a dendritic structure, specifically at the middle and top of the clad in sample B4. Sample B3 shows well developed TiC dendrites compared to that of sample B4.

Fig.14-c shows that sample B5 has fine TiC particles at the bottom of the clad (which gradually become coarser) and dendritic TiC in the middle. B5 microstructure shows both the dendritic and spherical morphology of TiC. The observed microstructure in B5 is not fully dendritic, similar to B3. However, dendrites of TiC can be observed in some locations in the middle and at top of the clad.

Fig.14-d depicts sample B6 with a *PDD* of 0.010. TiC particles are distributed uniformly in the clad. There is no evidence for the presence of a dendritic structure of TiC particles. TiC morphology in sample B6 is spherical and the TiC particle aspect ratio can be found between 1:1 and 1:5. As Table 4indicates, hardness values for this sample are high (e.g., 1300-3500 HVN).

This portion of the study on TiC morphology proves that *PDD* plays a crucial role in TiC formation and on the observed morphology of TiC particles. A considerable difference in TiC morphology can be detected for samples B1 and B2 which have identical laser parameters and *Eeff*, but different *PDD*, resulting from varying powder feed rate. The same can be said for the B3 to B6 series of data.

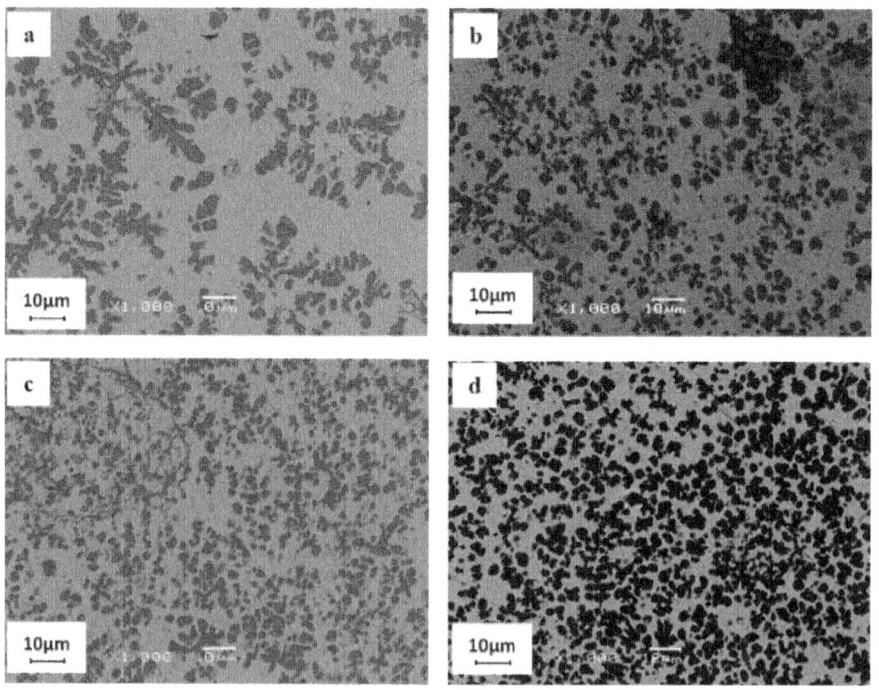

Figure 14: a) Sample B3 b) sample B4 c) sample B5 d) sample B6.

Constant Powder Deposition Density with Variable Effective Energy (Experiments Series C)

In this section, TiC particle morphology resulting from a constant powder deposition density with varying $Eeff$ was studied. For this purpose, a series of samples which are identical in PDD and different in $Eeff$ are selected. Samples C1 to C4 in the first group and C5 to C7 in the second group were investigated.

In the first group (C1 to C4), the scan speed and powder feed rate are constant. Therefore, laser power is the only parameter that is variable to create a different $Eeff$. All the samples show a dendritic structure for the TiC particles. An $Eeff$ range of 177 to 196 does not have a considerable effect on TiC morphology with a constant PDD (0.013). Table 4 shows little change in dilution for this series of samples. Since all the parameters (except laser power) are the same, it can be seen that in the range of 884 to 982 W, these parameters form a dendritic structure.

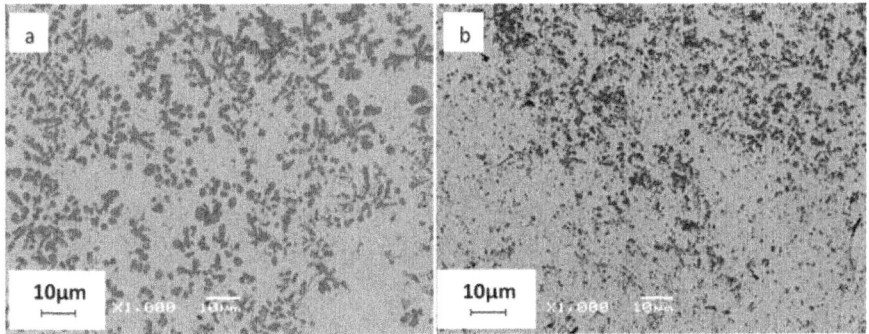

Figure 15: BSE image of TiC morphology resulting from different PDD a) sample C4 b) sample C7.

Thus, it may be concluded that with constant powder deposition density and laser parameters (such as scan speed, powder feed rate), increasing the $Eeff$ (i.e., increasing the laser power) likewise increases the probability of creating a liquid melt pool. As a result, the TiC particle structure is mostly dendritic. This interpretation is applicable to the first group of samples (C1 to C4).

Observing samples C5 and C7 provides valuable information. Other than for the laser power, they have the same laser parameters. Sample C5 has a lower laser power (700) than sample C7 (982). From the clad microstructures, it can be seen that dendritic TiC can be detected in sample C5 whereas sample C7 shows very fine TiC particles but no evidence of a dendritic structure. Higher laser power can cause increased dilution, which changes the clad composition to the Fe-rich side and precludes dendrite formation due to the small freezing range, as shown at the left side of the phase diagram.

Identical Effective Energy and Powder Deposition Density (Experiment Series D)

The purpose of this section is to clarify the effect of identical effective energy and powder deposition density on the morphology of the TiC particles formed during in-situ laser cladding. As mentioned earlier, the process parameters have been calculated to be in the high quality clad zone and also resulted in a specific effective energy and powder deposition density. Table 4 shows that sample D1 and D2 have the same effective energies ($Eeff$) and powder deposition densities (PDD), but dissimilar process parameters. The same situation can be seen in samples D3, D4, D5 and D6. Laser conditions of D1 and D2 have been used later in powder composition study (i.e. conditions A and B, respectively).

Fig.16-a illustrates back scatter images of sample D1 where dendritic TiC can clearly be seen. Conversely, Fig.16-b illustrates the TiCmorphology for sample D2, and no evidence of a dendritic structure for TiC can be found. Moreover, the distribution of TiC particles in the clad is very uniform compared to sample D1. A simple comparison between Fig.16-a and 17-b shows there is a considerable difference between TiC morphology, although both have the same $Eeff$ and PDD values. Both samples (D1, D2) show fine TiC at the clad bottoms which gradually becomes coarser at the top of the clad.

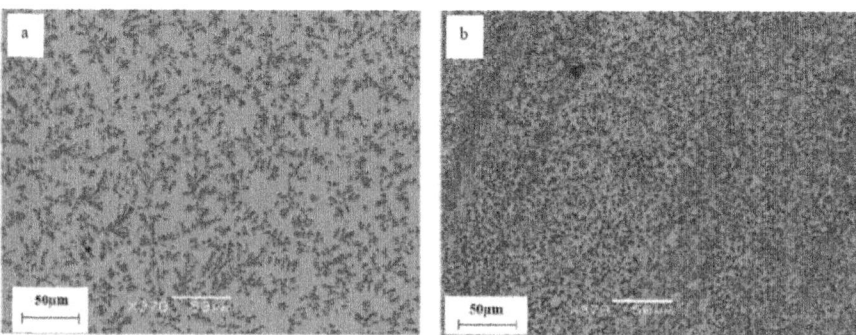

Figure 16: TiC morphology of samples a) D1 and b) D2.

DISCUSSION

In order to create a metallurgical bond between the clad and the substrate, dilution must occur. Fig. 9 shows that dilution by Fe decreases the liquidus temperature of the melt pool (labeled dot). Fig.10 illustrates a vertical section through the ternary Fe-Ti-C phase diagram at a constant composition of 19%at C using the Factsage software. According to the Fe-Ti-C ternary diagram, depending on the melt pool temperature (conditional on laser process parameters), ternary liquid can be present either in a single phase or in equilibrium with TiC/C to form a semi-solid melt pool. In a semi-solid melt pool, TiC and liquid are in equilibrium, as is shown in Fig.10. Further, a semi-solid melt pool resulted from a peak temperature less than the liquidus temperature, and spherical TiC particles which are distributed uniformly or in clusters can form from this mechanism.

In contrast, if the peak temperature exceeds the liquidus temperature, a liquid melt pool can be formed. Therefore, according to Figure 13a, TiC dendrites are the first phase to solidify during cooling and create dendritic TiC.

Fig.16 depicted different microstructures for samples D1 and D2. Sample D1 contains dendritic TiC, whereas sample D2 shows spherical TiC particles which are distributed in the matrix. Since the *Eeff* and *PDD* are identical for both, the only reason for the dissimilar microstructure could be the laser parameters, specifically the laser power. According to Fig. 9, dilution with Fe decreases the liquidus temperature. Peak power (i.e., 884) increases the melt pool temperature and thus increases the probability of having a liquid melt pool. Hence, the clad microstructure exhibits a well-developed dendritic structure of TiC.

In sample D6, although a higher laser power has been applied compared to sample D1 (i.e., 961), a dendritic TiC could not be detected. The diagram confirms that there is a large region in which liquid and solid TiC can be in equilibrium. This region decreases by increasing the Fe content. Therefore, in samples D4 and D6 (which have lower powder deposition density [i.e., 0.008]), dilution plays a vital role. The clad composition increases in Fe, and dendritic structures of TiC did not develop during solidification. On the other hand, Fig.10 shows that by increasing the Fe content by dilution, the freezing range decreases. By decreasing the freezing range, the chance of dendrite formation decreases. This phenomenon can explain the TiC morphology in C5 and C7. Dendrites of TiC were observed in C5 with a laser power of 700 W, whereas no evidence of dendritic TiC was observed in sample C7 with a laser power of 900 W.

In laser cladding, powder and substrate absorbed the energy in order to melt and create the clad. By increasing the powder feed rate, the energy portion absorbed by the powder increases, allowing for a decrease in laser energy absorbed by the substrate. As a result, dilution with the substrate is decreased. Increasing the laser power, or increasing the scan speed, increases the contribution of dilution (as seen in Table 4). By comparing samples D1 and

D2, it can be seen that dilution in sample D1 is higher than sample D2.

By increasing the dilution amount, the liquidus temperature decreases and shifts the clad composition to the Fe-rich side of the phase diagram. This results in a change in microstructure of the clad and TiC morphology.

Several modelling techniques have been used to model laser cladding by powder injection. These help to illustrate the physics of the process. A review of the attenuation of laser power in different powder feed rates and nozzle angles were presented, and the temperature distribution in the substrate and the powder particles on the substrate's surface was calculated (Huang, 2005).

POWDER COMPOSITION STUDY

In this section, the effect of powder composition on TiC morphology and clad hardness using the in-situ laser cladding process was studied. Two atomic ratios, 45:55 and 55:45, were selected for C:Ti (Fig. 1), the first one of which has the potential to form near stociometric TiC, while the second potentially forming residual graphite which may be beneficial from a wear perspective.. Fe percentages were explored with 70, 60, 50 and 10wt% to increase the volume fraction of TiC in the clad (Table 1). Table 5 shows the selected laser parameters for deposition as a function of composition. Note that, for samples 155, the powder feed rates were decreased, compared to the more Fe rich compositions, in order to achieve a bonded clad. Any feed rate higher than that reported inTable 5 did not create metallurgical bonding between the clad and substrate.

Table 5: Applied laser parameters

Parameters condition	Sample No	Laser power W	Scan speed (mm/sec)	Powder feed rate (g/min)	Effective Energy (J/mm2)	Powder deposition density (g/mm2)
A	745, 755,655,555	884.2	117	4	180	0.014
B	745, 755,655,555	663.4	88.5	3	180	0.014
A	155	884.2	117	2.12	180	0.007
B	155	663.4	88.5	0.95	180	0.004

SEM, EDS and Image Analysis Results

In this section, clad microstructures of samples with different clad compositions are studied. Fig.17shows the SEM images of samples with identical laser parameters but different chemical compositions. Results show that clad composition affects the clad microstructure and morphology. Fig.17-a and 18-bare related to the samples which are identical in Fe weight percentage but different in C:Ti ratio.Fig.17-a shows the dendiritic microstructure of TiC, while Fig.17-b depicts the spherical TiC distributed uniformly in the matrix. Decreasing the Fe percentage in the powder composition and increasing the C:Ti ratio increases the probability of a reaction between Ti and C and thus increases the volume fraction of TiC.

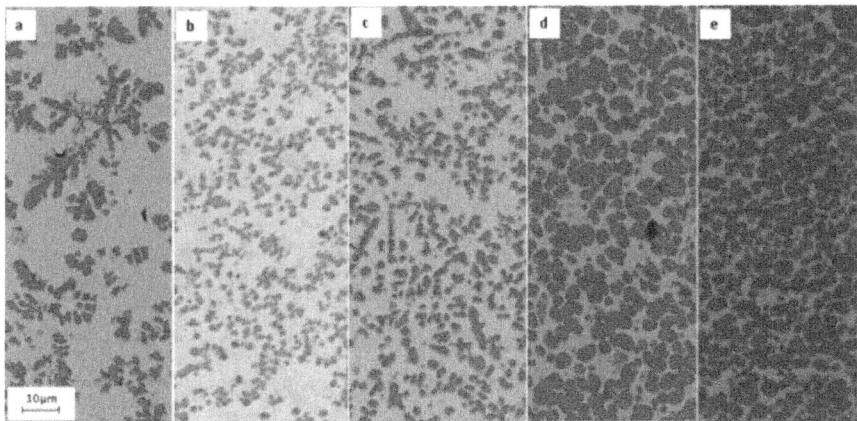

Figure 17: Developed TiC morphologies using laser condition A for compositions of: a) 745 b) 755 c) 655 d) 555 e) 155.

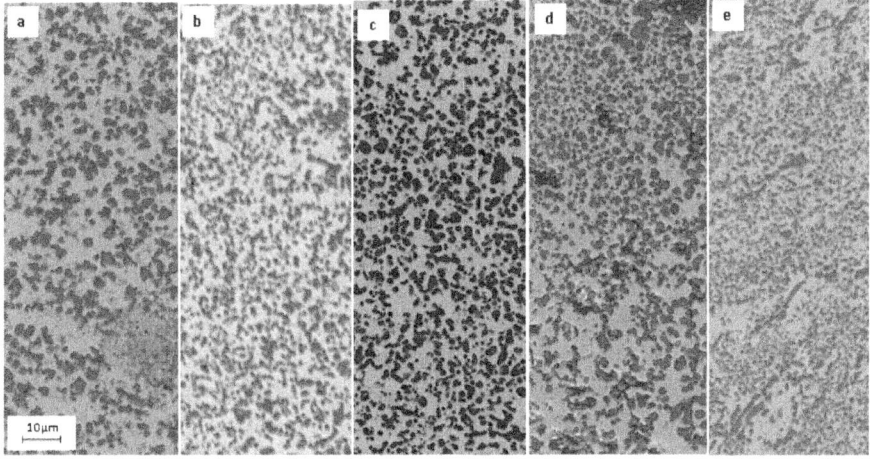

Figure 18: Developed TiC morphologies using laser condition B for compositions of: a) 745 b) 755 c) 655 d) 555 e) 155.

Fig.17-b to 18-d show a higher volume fraction of TiC distributed in the clad. Except for sample 745-A, dendritic TiC cannot be observed in the rest of the samples. This trend also can be observed in samples deposition using condition B, as indicated in Fig.18. Fig.17 and Fig.18 show that by decreasing the Fe percentage and increasing the C:Ti ratio, the volume fraction of TiC increases. Moreover, in condition B, dendritic TiC cannot be observed in sample 745-B and the size of the TiC particles are generally smaller for condition B compared to condition A. Image analysis helps to measure the trend of volume fraction qualitatively. Table 6 depicts the image analysis results. It can be seen that in similar applied laser conditions, the volume fraction of TiC increases by decreasing the Fe percentages. It is clear that decreasing the Fe content increases the TiC volume fraction but that this feature is more significant in laser condition A.

Table 6: Image analysis results

Sample	%Matrix	%TiC	Sample	%Matrix	%TiC
745-A	81	19	745-B	56	44
755-A	69	31	755-B	63	37
655-A	66	34	655-B	50	50
555-A	43	57	555-B	54	46
155-A	30	70	155-B	45	55

Hardness Results

Hardness results are depicted in

Fig.19, which provides information about samples with different chemical compositions deposited by laser condition A. Each hardness value is an average of at least three measurements at an identical position from the clad/substrate interface. It can be seen that, by decreasing the Fe percentage, general hardness values are increased. Sample 155-A with a minimum Fe percentage shows the highest clad hardness. This trend can also be seen in samples deposited by condition B.

Fluctuation in hardness results normally happens in composites because of different hardness values of matrix and reinforcement particles. Moreover, an average of three measurements can be a source of fluctuation depending on which phase is punched by the indenter. Sample 745-A with dendritic TiC has the lowest fraction of TiC particles (as illustrated in Fig.17-a and Table 6), causing a generally lower hardness compared to the other samples. Deposition of the same composition (sample 745-A) under different laser conditions (i.e., sample 745-B) increases the hardness profile dramatically. The main reason for hardness enhancement is the different distribution and morphology of TiC particles in 745-A compared to those in 745-B, which also significantly increases the TiC volume fraction from 19 to 44%.

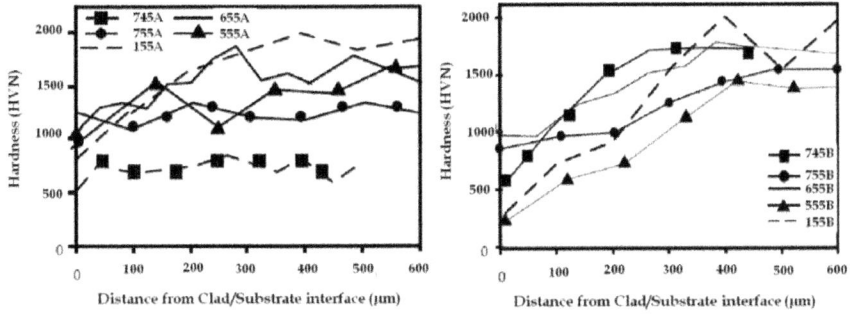

Figure 19: Hardness profile for samples deposited with laser condition A, B.

DISCUSSION

The temperature and composition of a clad melt pool during deposition are two main factors that affect final clad microstructure and TiC morphology. Laser power establishes the peak power that determines the clad temperature. Depending on the clad position, different cooling and solidification rates are developed which affect the clad microstructure and carbide morphologies. Composition is affected by initial fed powder composition and the extent of dilution by the Fe-rich matrix. Fig. 9 shows the liquidus projection in a ternary phase diagram of Fe-Ti-C. If no dilution by the substrate is assumed, then it is possible to state that the liquidus temperature of the powder mixture is about 2618 K (2345°C) in a powder composition of 745 and the liquidus temperature of composition 155 is about 3384 K (3111°C). Thermal modelling of A and B laser conditions shows peak temperatures of 2200°C and 1831°C, respectively (Emamian, 2010d). Only sample 745A exhibits a dendritic TiC morphology arising from solidification of a completely liquid melt pool. This indicates that the laser peak temperature exceeded the liquidus temperature of the clad composition. The main reason for the formation of a complete liquid melt pool in this sample is dilution by the substrate which shifts the clad composition to the Fe-rich side of Fig. 10, resulting in a decrease in the liquidus temperature. In the rest of the samples, as Fig. 9 shows, increasing the C:Ti ratio and decreasing the Fe percentage in the powder mixture, causes an increase in the liquidus temperature. Therefore, the peak temperature developed by the laser heating does not exceed the liquidus temperature for the clad composition.

Therefore a full liquid melt pool does not develop and spherical TiC particles are formed from the semi solid melt pool. While samples 745-A and755-A have very close chemical compositions, differences in liquidus temperatures result in a considerable change in TiC morphology.

WEAR

Wear resistance is an important function in the balance of properties of metal matrix composites. Wear starts from softer components which are in contact with counterfaces (i.e., pin, abrasive material or steel ball). Therefore, in metal matrix composites with hard particles, the matrix is worn in the preliminary stages of the wear process. Different mechanisms of matrix degrading such as plastic deformation, micromachining, crack propagation and brittle chipping, which are controlled by the matrix mechanical properties (i.e., hardness or fracture toughness) can increase the wear rate.

Wear can usually be classified on the basis of the nature of interactions, such as two-body or three-body. For two-body abrasive wear, a delamination model is feasible. In this model, wear resistance strongly depends on fracture toughness and not on composite strength, because crack propagation is the controlling factor. In three-body abrasive wear, abrasive particles tend to break down and bury themselves in alloys with relatively soft matrices. However, hard particles in composites resist scratching because of their high hardness under low loads and crack into small pieces under heavy loads. The broken particles remain embedded in the matrix during wear. In this case, composite strength and fracture toughness both play important roles (Jahanmir et al., 1976; Saka et al. 1985).

In this section, the wear resistances of six samples with different chemical composition deposited using two laser parameters (i.e., A, B) are studied. The objectives of this section are:

1. to find the optimal micro-structure (resulting from the laser process parameters and powder composition) of TiC in an Fe matrix with respect to the highest wear resistance ;

2. to measure the wear properties of in-situ deposited Fe-TiC by laser cladding; and

3. to find the optimal operating parameters of laser cladding that develop the best wear properties.

The wear resistanceof samples is measured based on standard ASTM G65-04 using an adjusted volume loss (mm^3), as below:

Volume loss (mm³) = weight loss (g)/density of coating (g.cm⁻³)*1000

$$(5)$$

Adjusted volume loss (mm³) = measured volume loss
  (mm³)*228.6 (mm)/wheel diameter after use (mm)

$$(6)$$

Results

In the previous section, Fig.17 and

Fig. 18 showed the TiC morphology resulting from different powder compositions and laser process parameters. In this section, wear properties of Fe-TiC composite coatings with different morphology are studied. Fig.17 and

Fig.18 show the distribution of TiC; however, the size of the particles and volume fraction are different.

The micro-hardness profile of conditions A and B show that samples 155A and 155B have a higher hardness value, which primarily have results from a higher volume fraction of the clad's TiC particles. In addition, sample 745A shows the lowest profile among the other samples. In considering Fig.17, it can be concluded that a larger area of the matrix remained without the support of TiC particles. Fig.20illustrates the wear test results based on ASTM G65-04.

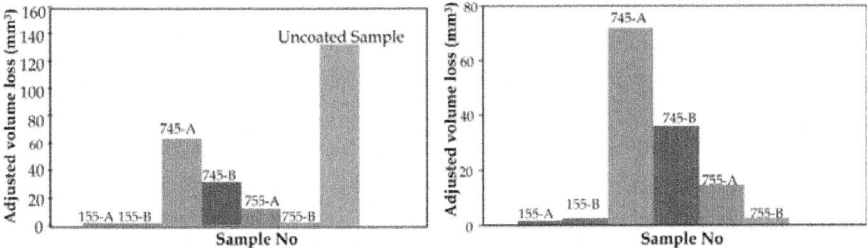

Figure 20: Wear test results.

The maximum volume loss after the uncoated sample (AISI 1030) belongs to sample 745A, while samples 155A and B have a minimum volume loss. Fig.20 shows that by increasing the C:Ti ratio in samples, volume loss decreases. By comparing samples 745 and 755 in both conditions A and B, group 755 shows better wear resistance. Moreover, although 755B has 70%wt Fe, the results show that it has good wear resistance compared to group 155. By comparing hardness and wear test results, it can be seen that samples with higher hardness show higher wear resistance.

DISCUSSION

By increasing the C:Ti ratio from 45:55 at% and 55:45 at% with the same laser condition, the TiC particles' morphology was formed into spherical shapes with a uniform distribution. Group 155A has a higher volume fraction of TiC compared to the 155B conditions (Table 6). By considering the melting points of ingredients (3400, 1670, and 1535°C for C, Ti and Fe, respectively), higher amounts of C and Ti in group 155 absorb a considerable portion of the laser energy compared to 745 or 755 group composition. Hence, the higher laser power in sample 155A compared to 155B encourages a reaction between C and molten Ti by the quick melting of Ti and Fe. However, in 155B, the laser energy is not sufficient to promote TiC formation in the clad. By comparing Fig.17-c and

Fig. 18-c, it can be seen that the volume fraction and size of the TiC particles are larger in 155A than in 155B, resulting in less volume loss in 155A. Also, the larger TiC particle size in 155A compared to 155B can be another reason for better wear resistance in 155A than in 155B.

In sample 755A, a higher laser beam energy melts the substrate and increases the dilution. Hence, the volume fraction of the TiC decreases compared to 755B. Therefore, laser parameters can be selected based on powder composition in order to have a maximum TiC volume fraction.

Fig.20 depicts how the abrasive wear performance of samples is affected by the volume fraction of TiC particles. The volume fraction of TiC particles in reverse is related to the distance between carbides occupied by the Fe matrix. It is obvious that, in abrasive conditions, the softer part is removed by grits. Therefore, it is expected that by increasing the volume fraction of the coating, the wear resistance increases (He et al., 2009; Nurminen et al., 2009; Pu, 2008). The volume fraction in this research is increased firstly by increasing the C:Ti ratio in the constant Fe percentage and secondly by decreasing the Fe percentage in the powder composition. Optimal results for wear resistance of in-situ Fe-TiC can be found in group 155A. Since Ti can dissolve in Fe during the laser cladding process, nominal 90% TiC (93%vol TiC) is not achieved during the process; however, maximum 70% volume fraction of TiC shows considerable wear resistance.

The usual mechanism of abrasives in wear tests are grooving, plastic deformation and fracture of hard phases. Table 7 provides information about size and hardness values of hard-phase particles and abrasives.

Table 7: Particle size information.

Particle	Size range, μm	Hardness (HVN)
TiC	1-7	3200
Abrasive	212-300	1160

Abrasive grits first start deforming the Fe matrix, which is the softer component in the composite coating. If the carbide interlock is sufficiently high, the exposure of the soft matrix to the abrasive will be decreased. Since the hardness of the TiC is higher than abrasives (Table 7), carbide damage is less than for grits. Moreover, in-situ TiC particles are developed during the laser cladding process, and hence the TiC/Fe interface has a strong bonding. Therefore, the probability of deboning and crack formation in the particle/binder interface decreases. Pulling out and crushing the WC particles are the main problems that have been noted by researchers (Stachowiak, G. B. & Stachowiak G.W, 2010). Furthermore, identical wear test conditions were conducted on sample WC-12%Ni (with 80%vol WC) to compare it with 155A (with 80% vol TiC). Results show a similar adjusted volume loss for both (i.e., 1.9744 and 1.7731 mm³ volume loss for WC and 155A, respectively).

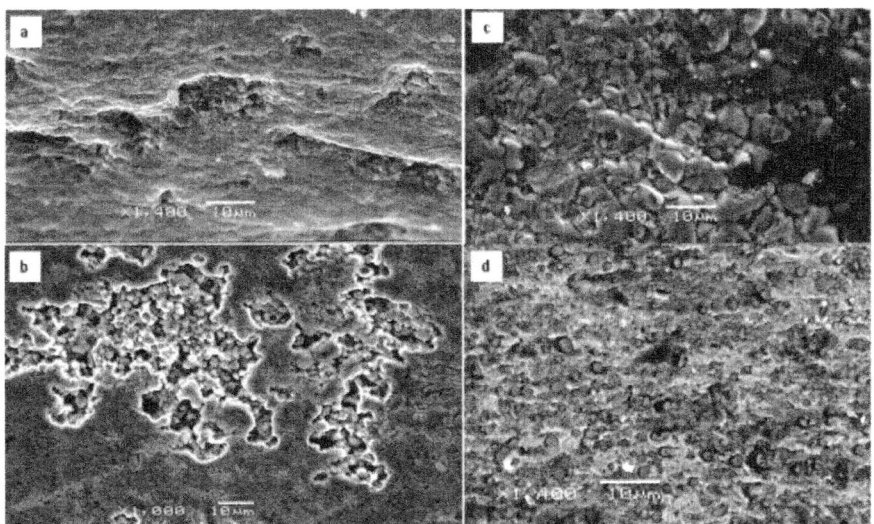

Figure 21: a) Plastic deformations in worn surface of 745A (b) Pulling out of carbides in worn surface of 755A (c) Worn matrix and remaining TiC particles in sample 155A (d) Worn matrix and remaining TiC particles in sample 755B.

It is normal that the matrix of sample 745A, which has minimum support of the TiC, contains more plastic deformation than the others (Fig. 21 a). Matrix plastic deformation increases the stress on TiC particles, pulling them off the surface. It is worth noting that, during the wear test, increasing the temperature in the matrix around the TiC particles helps to drop the yield stress of the matrix, resulting in easier plastic deformation around the carbides. By increasing the volume fraction of the TiC in samples 755 and 155, plastic deformation could not be observed. However, in some areas of 755A, a pulling out of the TiC is detected. Fig. 21 b shows a pulling out of the carbides in sample 755A. Group 155 shows a uniform worn surface without carbide removal. The higher volume fraction of the TiC results in the minimizing of the plastic deformation and removing of the matrix. However, volume loss still belongs to the matrix area. Fig. 21-c shows the interlocked TiC and a small area of worn matrix in 155A. Fig. 21-d depicts the worn matrix for sample 755B with the remains of the TiC. Although sample 755B contains 70% Fe, the uniform distributed particles and high volume fraction of the TiC resulted in higher wear resistance than that of 755A and 745A and B.

CONCLUSION

From the above experiments, we can conclude that two important factors – temperature and chemical composition of the clad – establish clad microstructures. We can also conclude that laser process parameters play a crucial role in the quality and microstructure of the clad, the ternary phase diagram of Fe-Ti-C can be used to predict the final microstructure of the clad. Furthermore, the melt pool chemical composition defines the liquidus temperature. Hence, the clad temperature can be higher or less than the liquidus temperature in different locations of the clad. Therefore, depending on clad and liquidus temperatures, the melt pool would be semi-solid or liquid, which defines the TiC formation mechanism and morphology.

The experiments also showed that, depending on powder deposition density (resulting from the powder feed rate and scan speed) and peak power of the clad temperature, different TiC morphologies and distributions could result. Moreover, identical combined parameters do not guarantee identical TiC morphology, as laser parameters still play a role in TiC morphology.

Furthermore, a uniform distribution of spherical TiC particles indicates that increasing the fraction of carbides in a matrix results in higher values for hardness (sample D2). From this, it can be concluded that, in order to enhance the clad hardness and wear resistance and achieve uniform carbide distribution, laser parameters must be adjusted to form a semi-solid melt pool instead of a liquid molten pool. A liquid melt pool forms dendritic TiC particles, which are not well distributed through the clad. Hence, a considerable region of clad cannot be supported by TiC particles. This conclusion also was confirmed by wear resistance results in a series of chemical composition and two laser conditions.

REFERENCES

1. S. Ariely, J. Shen, M. Bamberger, F. Dausiger, H. Hugel, 1991 Laser surface alloying of steel with TiC. Surface and Coatings Technology,45403 .

2. C. Cui, Z. Guo, H. Wang, J. Hu, 2007 In situ TiC particles reinforced grey cast iron composite fabricated by laser cladding of Ni-Ti-C system. Journal of Materials Processing Technology, 183 380385 .

3. K. Das, T. K. Bandyopadhyay, S. Das, 2002 A review on the various synthesis routes of TiC reinforced ferrous based composites. Journal of Materials Science 37 38813892 .

4. L. Dubourg, L. St-Georges, 2006 Optimization of laser cladding process using Taguchi & EM methods for MMC coating production. Journal of Thermal Spray Technology 15 790795

5. B. Du, Q. Li, X. Wang, Z. Zou, 2007 In situ synthesis of TiC-VC particles reinforced Fe-based metal matrix composite coating by laser cladding. Hanjie Xuebao/Transactions of the China Welding Institution 28 6568 .

6. B. Du, Z. Zou, X. Wang, Q. Li, 2007 In situ synthesis of TiC-TiB2 reinforced FeCrSiB composite coating by laser cladding. Surface Review and Letters 14 315319 .

7. A. Emamian, S. F. Corbin, A. Khajepour, 2010 Effect of laser cladding process parameters on clad quality and in-situ formed microstructure of Fe-TiC composite coatings. Surface and Coatings Technology 205 20072015 .

8. A. Emamian, S. F. Corbin, A. Khajepour, 2010 "Study on Laser Parameters Effect on Morphology of In-Situ Fe-TiC particles Deposition on Mild Steel Using Laser Cladding Process" ICALEO Conference 2010, Anheim, California, USA.

9. A. Emamian, S. F. Corbin, A. Khajepour, 2010 "Correlation between temperature distribution and formed microstructure of in-situ laser cladding of Fe-TiC on carbon steel" ICALEO Conference 2010, Anheim, California, USA.

10. A. Emamian, S. F. Corbin, A. Khajepour, 2009 "In-situ TiC particles reinforced carbon steel composite Fabricated by laser cladding of Fe-Ti-C system", 21st Canadian Material Science Conference, Kingston, Canada.

11. Q. He, Y. Wang, W. Zhao, Y. Cheng, 2009 Interface microstructure and wear properties of TiC-Ni-Mo coatings prepared by in-situ fabrication of laser cladding. Hanjie Xuebao/Transactions of the China Welding Institution 30.

12. Okamoto. Hirokai, 2000 Phase Diagrams for Binary Alloys (ASM)

13. Y. L. Huang, 2005 Interaction between laser beam and powder stream in the process of laser cladding with powder feeding. Journal of Modeling and simulation in material science and engineering 13 4756 .

14. S. Jahanmir, E. P. Abrahamson, N. P. Suh, 1976 Wear, 40 75

15. W. H. Jiang, J. Fei, X. L. Han, 2000 In situ synthesis of (TiW)C/Fe composites. Materials Letters, 46 222224 .

16. W. H. Jiang, R. Kovacevic, 2007 Laser deposited TiC/H13 tool steel composite coatings and their erosion resistance. Journal of Materials Processing Technology, 186 331338 .

17. J. Nurminen, J. Näkki, P. Vuoristo, 2009 Microstructure and properties of hard and wear resistant MMC coatings deposited by laser cladding. International Journal of Refractory Metals and Hard Materials 27 472478 .

18. Y. Pu, 2008 Microstructure and tribological properties of in situ synthesized TiC, TiN, and SiC reinforced Ti3Al intermetallic matrix composite coatings on pure Ti by laser cladding. Appl. Surf. Sci. 255 26972703 .

19. N. Saka, D. P. Karalekas, 1985 Friction and Wear of Particle Reinforced Metal-Ceramic Composites, Wear of Materials, 784

20. G. B. Stachowiak, G. W. Stachowiak, 2010 Tribological characteristics of WC-based claddings using a ball-cratering method. International Journal of Refractory Metals and Hard Materials 2895 .

21. C. Tassin, F. Laroudie, M. Pons, L. Lelait, 1995 Carbide-reinforced coatings on AISI 316 L stainless steel by laser surface alloying. Surface and Coatings Technology, 76-77, 450 EOF455 EOF .

22. X. H. Wang, S. L. Song, S. Y. Qu, Z. D. Zou, 2007 Characterization of in situ synthesized TiC particle reinforced Fe-based composite coatings produced by multi-pass overlapping GTAW melting process. Surface & Coatings Technology,201 58995905 .

23. X. H. Wang, M. Zhang, X. M. Liu, S. Y. Qu, Z. D. Zou, 2008 Microstructure and wear properties of TiC/FeCrBSi surface composite coating prepared by laser cladding. Surface and Coatings Technology, 202 36003606 .

24. X. H. Wang, S. Y. Qu, B. S. Du, Z. D. Zou, 2009 In situ synthesised TiC particles reinforced Fe based composite coating produced by laser cladding. Materials Science and Technology 25 388392 .

25. X. Wu, Y. Hong, 2001 Microstructure and mechanical properties at TiCp/ Ni-alloy interfaces in laser-synthesized coatings. Materials Science and Engineering A 318 1521 .

26. M. Yan, H. Hanqi, 1996 In situ laser surface coating of TiC metal-matrix composite layer. J. Mater. Sci. 31 43034306 .

Chapter 9

POLYPROPYLENE/GRAPHENE AND POLYPROPYLENE/CARBON FIBER CONDUCTIVE COMPOSITES: MECHANICAL, CRYSTALLIZATION AND ELECTROMAGNETIC PROPERTIES

Chien-Lin Huang [1], Ching-Wen Lou [2], Chi-Fan Liu [3], Chen-Hung Huang [4], Xiao-Min Song [5] and Jia-Horng Lin [5,6,7]

[1]Department of Fiber and Composite Materials, Feng Chia University, Taichung City 40724, Taiwan

[2]Institute of Biomedical Engineering and Materials Science, Central Taiwan University of Science and Technology, Taichung City 40601, Taiwan

[3]Office of Physical Education and Sports Affairs, Feng Chia University, Taichung 407, Taiwan

[4]Department of Aerospace and Systems Engineering, Feng Chia University, Taichung City 40724, Taiwan

[5]Laboratory of Fiber Application and Manufacturing, Department of Fiber and Composite Materials, Feng Chia University, Taichung City 40724, Taiwan

[6]School of Chinese Medicine, China Medical University, Taichung City 40402, Taiwan

[7]Department of Fashion Design, Asia University, Taichung City 41354, Taiwan

ABSTRACT

This study aims to examine the properties of composites that different carbon materials with different measurements can reinforce. Using a melt compounding method, this study combines polypropylene (PP) and graphene nano-sheets (GNs) or carbon fiber (CF) to make PP/GNs and PP/CF conductive composites, respectively. The DSC results and optical microscopic observation show that both GNs and CF enable PP to crystalize at a high temperature. The tensile modulus of PP/GNs and PP/CF conductive composites remarkably increases as a result of the increasing content of conductive fillers. The tensile strength of the PP/GNs conductive composites is inversely proportional to the loading level of GNs. Containing 20 wt% of GNs, the PP/GNs conductive composites have an optimal conductivity of 0.36 S/m and an optimal EMI SE of 13 dB. PP/CF conductive composites have an optimal conductivity of 10^{-6} S/m when composed of no less than 3 wt% of CF, and an optimal EMI SE of 25 dB when composed of 20 wt% of CF.

INTRODUCTION

An intensively explored subject [1,2,3,4,5,6,7,8,9,10,11,12], polymer conductive composites are functional composites made by adding conductive fillers to polymers using a specified processing method, yielding steady and sustained electrical conductivity. Their electrical conductivity can also be adjusted with a greater range, and they are easily processed. Due to the aforementioned advantages, polymer conductive composites are commonly used in diverse fields, such as electronics, energy sources, and chemical engineering.

The electrical conductivity of polymer conductive composites is greatly dependent on the properties of the polymer as well as the conductive fillers' type, content, geometrical shape, and dispersion. The three major categories of conductive fillers are carbon, metal and metallic oxide, the former of which includes carbon black (CB) [5,13,14,15,16], carbon fibers (CF) [1,5,15,17,18,19], graphite [18,20,21], carbon nanotubes (CNT) [2,4,14,22,23], and grapheme [4,24,25,26,27] and is the most frequently used due to its light weight, its easy formation of conductive networks, and its oxidation resistance. Moreover, polymer conductive composites commonly use polypropylene (PP) [1,5,16,20], polyethylene (PE) [1,18], and polystyrene (PS) [3,13,16,24] as their matrices. PP is a highly commercially available polymer due to its excellent mechanical properties, good heat resistance, low cost, ease of processing, and full recyclability.

A melt compounding method is more appealing than an in-situ polymerization method or solution mixing method, and its combination of traditional facilities, such as an extruder and a mixer, can give the production a greater diversity of polymers and fillers. Such a method is relatively economical and suitable for mass production [28] and has already been successfully applied in the production of polymer conductive composites with the combination of conductive fillers of CNT, CB, expanded graphite, and grapheme [1,2,14,16,18]. With a melt compounding method, PP (*i.e.*, the matrix) is combined with GNs or CF (*i.e.*, conductive fillers) to form PP/GNs and PP/CF conductive composites. The crystallization properties, tensile properties, conductivity and EMI SE of the conductive composites are characterized by DSC, optical microscope observation, tensile test, conductivity test, and EMI SE test.

EXPERIMENTAL SECTION

Materials

PP (1080, Formosa Plastics Corporation, Taiwan) is a homopolymer with a density of 0.9 g/cm³ and melt flow index of 10 g/10 min. CF (HTS 40, Toray Industries Inc., Tokyo, Japan) has a length of 6.2 mm and a diameter of 7 µm, as indicated inFigure 1a. GNs (Enerage Inc., Taiwan) have a thickness of 50–100 nm and a conductivity greater than 700 S/cm. The atomic force microscope (AFM) of GNs is indicated in Figure 1b.

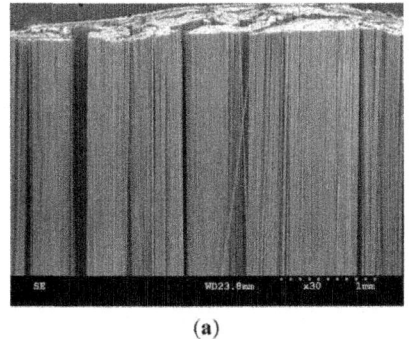
(a)

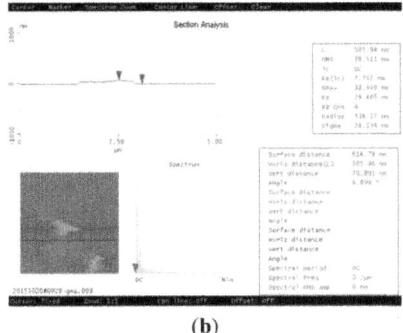
(b)

Figure 1: (a) SEM image of CF (carbon fibers) and (b) AFM image of GNs (graphene nano-sheets).

Methods

PP/GNs and PP/CF conductive composites were made using a melt compounding method by a Brabender mixer with a processing temperature set to 180 °C, a processing time of 5 min, and a screw speed of 75 rpm. Finally, a hot press machine processed the conductive composites at 200 °C for 5 min to give them a 0.5-mm thickness, and cooled them at room temperature.

Tests

The conductive composites were evaluated by a differential scanning calorimeter (DSC), tensile test, flexural test, optical microscope observation, conductivity test, EMI SE test. In terms of the crystallization properties, DSC measurement used a Q200 (TA Instruments, New Castle, DE, USA) with samples that were heated from 40 °C to 200 °C at 10 °C/min increments, after which samples stayed at 200 °C for 5 min, then were cooled from 200 °C to 40 °C, and heated from 40 °C to 200 °C at the same increments. In terms of the mechanical properties, the tensile test was conducted according to ASTM

D638. The optical microscopic observation was performed as follows. Tensile tests were performed via an Instron 5566 universal tester (Instron, Canton, MA, USA) as indicated in Figure 2, according to the specifications of ASTM D638-10. Samples were made into a dumbbell shape as seen in Figure 2c according to ASTM D638 Type IV. The crosshead speed was 5 mm/min and a total of five samples for each specification were used. A glass cover extended the samples into euphotic film at 200 °C on a specimen heating holder, and then each was placed in the optical microscope, isothermally cooled to 130 °C at 10 °C/min increments and kept at 130 °C so as to observe the crystallinity of PP. The conductivity of the conductive composites was evaluated by the four-point probe method, while the EMI SE of the samples was measured at a frequency range of 300 MHz to 3 GHz as specified in ASTM D4935.

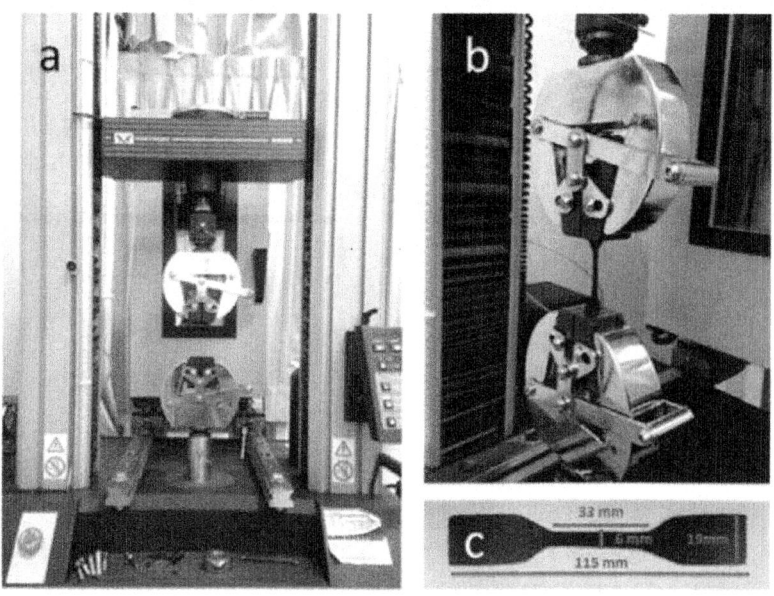

Figure 2: (**a**) The tester, (**b**) the grips, and (**c**) the samples of tensile tests.

RESULTS AND DISCUSSION

Performance of Crystallization

Table 1 and Table 2 show that PP has an initial crystallinity temperature (T_{onset}) of 116.21 °C and crystallinity temperature (T_c) of 111.46 °C, both of which shift to the high temperature zone as a result of the combination of GNs and also continuously increase as a result of the increasing content of GNs. The optimal T_{onset} of 140.67 °C and T_c of 135.01 °C are present when the PP/GNs conductive composites consist of 20 wt% of GNs, and are greater than those of pure PP by 24.46 °C and 23.55 °C, respectively. Similarly, T_{onset} and T_c of PP/CF conductive composites also shift to the high temperature zone as a result of the combination of CF, and also increase with the increasing levels of CF. When containing 20 wt% CF, PP/CF conductive composites have optimal T_{onset} (126.25 °C) and T_c (122.28 °C), which are greater than those of pure PP by 10.04 °C and 10.82 °C, respectively. Such results are ascribed to the combination of GNs or CF as the nucleating agent of PP, which makes the nucleation mode heterogeneous instead of homogeneous, significantly decreases the nucleation free energy, and allows the molecular chain to attach to and be arranged orderly on the nucleating agent [22,23]. GNs demonstrate a more significant influence over the T_{onset} and T_c of PP than CF does. Such a result is ascribed to the nanometer size effect of GNs, namely that the size of GNs is smaller than that of CF. With the same content, GNs contribute to a greater superficial area and at the same time greater nucleation sites, thereby demonstrating a greater effect on the T_{onset} and T_c of PP.

PP has $T_{onset}-T_c$ of 4.75 °C and $t_{1/2}$ of 0.444 min. The combination of GNs results in a decrease in crystallization rate of PP, exemplified by an increase in $T_{onse}-T_c$ of 4.79 °C to 6.1 °C and an increase in $t_{1/2}$ of 0.477 min to 0.604 min. By contrast, 15 wt% of GNs makes for the slowest crystallization rate of PP, making $T_{onset}-T_c$ 6.10 °C and $t_{1/2}$ 0.604 min. Conversely, the combination of CF accelerates the crystallization rate of PP, causing $T_{onset}-T_c$ to decrease to 3.70 °C–4.33 °C and $t_{1/2}$ to decrease to 0.358 min–0.443 min. In particular, 5 wt% of CF results in the shortest crystallization rate of PP, in which $T_{onset}-T_c$ is 3.70 °C and $t_{1/2}$ is 0.358 min. The combination of inorganic fillers to polymers has two influences over the crystallization behavior of the polymer. Using inorganic fillers as nucleating agent allows the molecular chains of PP to nucleate and the spherocrystals to enlarge at a high temperature, and at the same time allows for more nucleation sites [24,25,26,27,28,29]. On the other hand, the existence of inorganic fillers takes up a certain space, especially with nano-scale fillers, and restricts the motion and arrangement of molecular chains that influence the crystallization rate of the polymer. As a result, the crystallization rate of the polymer is inversely proportional to the inorganic fillers, but the combination of CF is proportional to the crystallization rate of PP.

The degree of crystallinity (X_c) of PP decreases slightly when the combination of GNs is below 5 wt%, and slightly increases when the combination of GNs is above 5 wt%. Conversely, the combination of CF hardly influences the X_c of PP.

Table 1: Crystallization properties of PP/GNs composites

wt%	T_c (°C)	T_{onset} (°C)	$T_{onset}-T_c$ (°C)	$t_{1/2}$ (min)	X_c (%)
0	111.46	116.21	4.75	0.444	40.3
1	124.91	129.70	4.79	0.479	38.6
3	127.54	132.43	4.89	0.477	38.6
5	129.39	135.22	5.83	0.567	45.1
10	132.75	138.19	5.44	0.543	44.5
15	133.36	139.46	6.10	0.604	45.7
20	135.01	140.67	5.66	0.574	45.4

Table 2: Crystallization properties of PP/CF composites.

wt%	T_c (°C)	T_{onset} (°C)	$T_{onset}-T_c$ (°C)	$t_{1/2}$ (min)	X_c (%)
0	111.46	116.21	4.75	0.444	40.3
1	118.92	122.73	3.81	0.373	38.9
3	119.42	123.27	3.85	0.368	39.6
5	120.23	123.93	3.70	0.358	39.9
10	120.61	124.56	3.95	0.403	40.4
15	120.90	125.23	4.33	0.443	41.8
20	122.28	126.25	3.97	0.387	41.1

Figure 3 shows the image series occurring during the process that GNs are added to PP, and cooled from a melted state at 200 °C at 10 °C/min increments until it reaches 130 °C, where it is kept. The presence of GNs can be seen in the melted state image in Figure 3a. Figure 3b is the image photographed close to 130 °C, at which point the molecules start to nucleate and PP starts to nucleate the surrounding GNs. Figure 3c,d show that the spherocrystals continuously enlarge. Due to the nano-scale that GNs are at, the combination of GNs results in numerous nucleating sites in PP. Therefore, there is a great amount of spherocrystals of PP, which easily collide with adjacent spherocrystals, and as a result, spherocrystals have a small size and low completeness.

Figure 4 shows the image series occurring during the process in which CF was added to PP and cooled from a melted state at 200 °C at 10 °C/min increments until it reached 130 °C, where it was kept. The black part shown in Figure 4a is CF;Figure 4b shows that the molecular chains of PP nucleate and the spherocrystals enlarge along the CF; Figure 4c,d shows ever-growing spherocrystals and a clear transcrystallization. Compared to the combination of GNs, the combination of CF makes spherocrystals larger and more complete.

Figure 3 and Figure 4 show that using GNs or CF as the nucleating agent facilitates the crystallization of PP. The combination of GNs results in a smaller spherocrystal size but a poorer completeness. Conversely, the combination of CF results in a larger spherocrystal size but a better completeness.

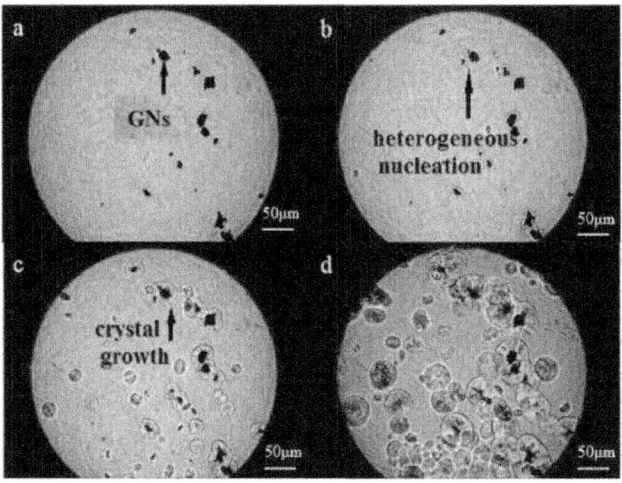

Figure 3: Optical microscopic images (500×) of PP/GNs composites at (**a**) 200 °C and (**b–d**) near 130 °C.

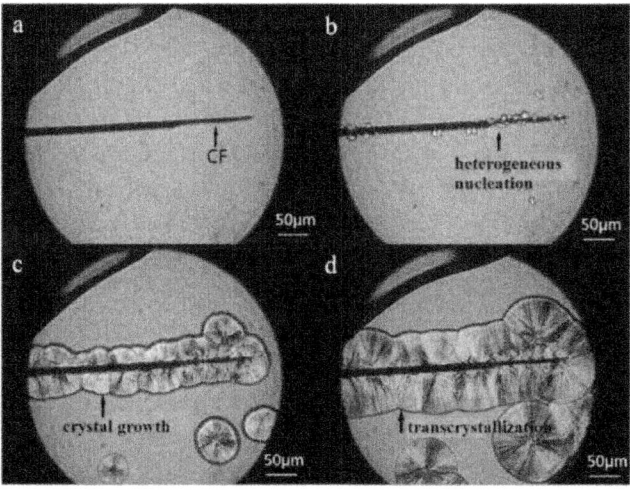

Figure 4: Optical microscopic images (500×) of PP/CF composites at (**a**) 200 °C and (**b–d**) near 130 °C.

Performance of Tensile Properties

The mechanical properties of polymer composites are highly correlated with the intrinsic properties, amount, dispersion of the fillers, the properties of the polymer matrix, and the interaction between fillers and polymer matrix. Figure

5 andTable 3 indicate the significant influence of the loading level of GNs over the mechanical properties of PP/GNs conductive composites and show that it is inversely proportional to the tensile strength but proportional to the tensile modulus. Compared to the mechanical properties of pure PP, the PP/GNs conductive composites containing 20 wt% of GNs have a 34% lower tensile strength (from 33.74 MPa to 22.28 MPa) and a 150% greater tensile modulus (from 1621 MPa *vs.* 4037 MPa).

In the tensile test, the fracture mode of PP is ductile fracture, which is quite different from that of PP/GNs conductive composites, which have an obvious yield phenomenon before fracture. Furthermore, when containing more content of GNs, PP/GNs conductive composites exhibit a classical brittle fracture, namely they do not exhibit a significant yield phenomenon before fracture. Additionally, their elongation at break is considerably low [30,31] as indicated in Figure 7a. GNs create a discontinuity of PP in PP/GNs conductive composites, and at the same time create a stress concentration phenomenon. When the composites are posed with load, the initial cracking occurs and then becomes worse where the stress concentration occurs, and eventually causes structural failure. Although serving as the dispersed phase, GNs are actually isolated from the PP matrix in the PP continuous phase, and resemble a sea-island structure. An increasing content of GNs inevitably decreases the force bearing area of matrix, thereby decreasing the tensile strength of the composites in comparison to that of pure PP. In addition, the conductivity of GNs stems from their own sp^2 hybridization structure; therefore, adding functional groups to GNs certainly destroys such a structure and then decreases GNs' initial conductivity. This study does not functionally modify GNs as it aims to develop polymer conductive composites with better conductivity. The modulus of the polymer composites can be augmented by adding fillers that have a high modulus to polymers that have a lower modulus, regardless of the interaction between them. Such a result is ascribed to a complex explanation, which is beyond any complete theories but it can be simply put that the fillers possess a restrictive effect over the polymers by restricting the motion and deformation of their molecular chains.

Figure 6 and Table 3 show that the increasing content of CF significantly influences the mechanical properties of the PP/CF conductive composites with their tensile strength first decreasing then increasing and the tensile modulus constantly increasing. Compared to the tensile strength (33.74 MPa) and modulus (1621 MPa) of pure PP, the combination of 20 wt% of CF causes the tensile strength of the PP/CF conductive composites to decrease by 8.3% to 30.94 MPa and increases their tensile modulus by 159% to 4207 MPa.

The combination of CF has a similar augmentation of the modulus to the combination of GNs. The PP/CF conductive composites do not exhibit an obvious yield phenomenon before fracture (*i.e.*, classical brittle fracture), and also have a remarkably low elongation at break [30,31], as indicated in Figure 7b. However, this elongation is still relatively greater than that of PP/GNs conductive composites. The increase in tensile modulus and the decrease in tensile strength of PP/CF conductive composites are caused by the same factors as that for PP/GNs conductive composites. When the content of CF is more than 3 wt%, the tensile strength of PP/CF conductive composites continuously increases, the result of which is ascribed to the cross structure caused by PP and a high content of CF. Some fibers are entangled and then provide the reinforcing effect, which is conducive to the tensile strength.

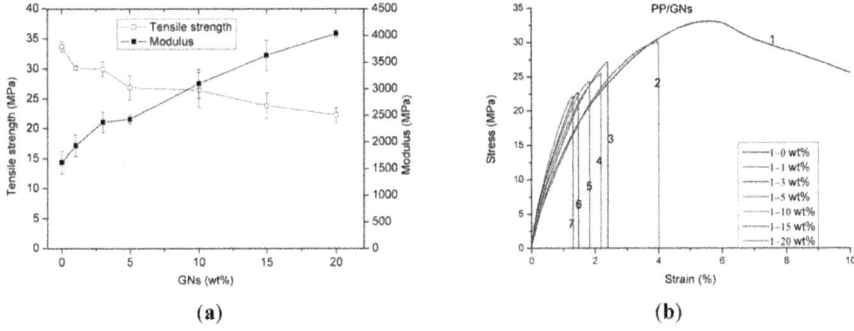

Figure 5: The tensile properties of PP/GNs composites: (**a**) tensile strength and modulus and (**b**) stress-strain curve.

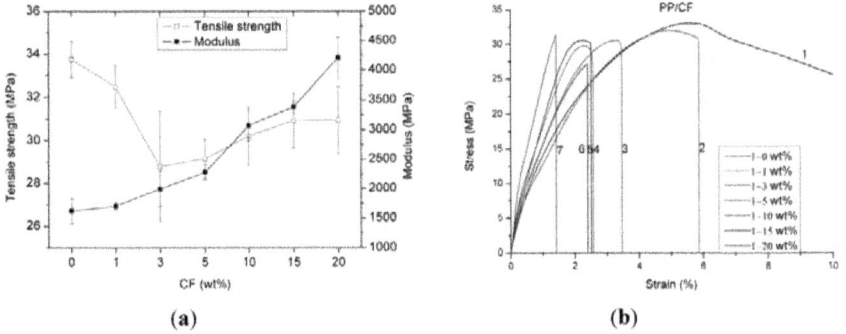

Figure 6: The tensile properties of PP/CF composites: (**a**) tensile strength and modulus and (**b**) stress-strain curve.

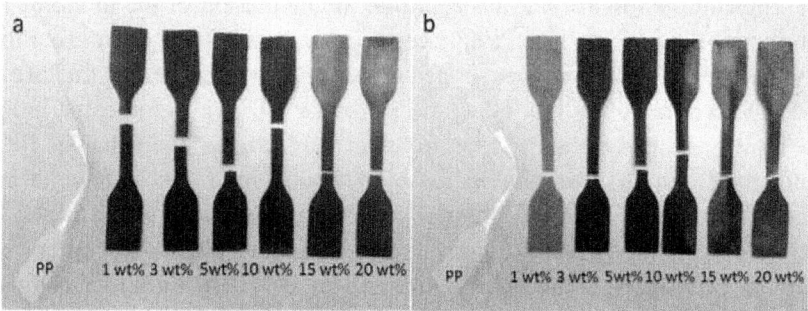

Figure 7: Stereomicroscopic images of the fractured samples: **(a)** PP/GNs composite and **(b)** PP/CF composites.

Table 3: Tensile properties and electrical conductivity of PP/GNs and PP/CF composites

Filler (wt%)	GNs				CF			
	Tensile strength (MPa)	Tensile Modulus (MPa)	Elongation (%)	Conductivity (S/m)	Tensile strength (MPa)	Tensile Modulus (MPa)	Elongation (%)	Conductivity (S/m)
0	33.75 ± 0.87	1621 ± 211	>100	$1.2 \times 10^{-13} \pm 1.1 \times 10^{-13}$	33.75 ± 0.87	1621 ± 211	>100	$1.2 \times 10^{-13} \pm 1.1 \times 10^{-13}$
1	30.16 ± 0.34	1934 ± 203	3.66 ± 0.75	$9.2 \times 10^{-10} \pm 2.4 \times 10^{-10}$	32.49 ± 0.98	1703 ± 60	5.75 ± 0.5	$3.4 \times 10^{-6} \pm 1.2 \times 10^{-6}$
3	29.98 ± 1.29	2374 ± 192	3.07 ± 0.33	$6.8 \times 10^{-10} \pm 1.3 \times 10^{-10}$	28.78 ± 2.59	1993 ± 301	2.64 ± 0.36	$3.8 \times 10^{-6} \pm 1.4 \times 10^{-6}$
5	26.92 ± 2.03	2429 ± 78	2.63 ± 0.39	$2.3 \times 10^{-5} \pm 1.2 \times 10^{-5}$	29.15 ± 0.91	2278 ± 129	3.54 ± 0.26	$3.8 \times 10^{-6} \pm 1.3 \times 10^{-6}$
10	26.46 ± 2.89	3099 ± 270	1.86 ± 0.15	$1.1 \times 10^{-2} \pm 0.003$	30.22 ± 1.33	3072 ± 252	2.59 ± 0.16	$3.1 \times 10^{-6} \pm 1.1 \times 10^{-6}$
15	23.88 ± 2.17	3630 ± 280	1.54 ± 0.16	0.12 ± 0.01	30.94 ± 1.27	3384 ± 102	2.57 ± 013	$3.2 \times 10^{-6} \pm 1.2 \times 10^{-6}$
20	22.28 ± 1.23	4037 ± 64	1.21 ± 0.21	0.36 ± 0.12	30.95 ± 1.56	4207 ± 357	1.62 ± 0.23	$3.2 \times 10^{-6} \pm 1.3 \times 10^{-6}$

Performance of Electrical Conductivity and EMI SE

Figure 8 shows that the conductivity of pure PP is 10^{-13} S/m (*i.e.*, the orders of magnitude), which shows the classical insulating property of polymer. The combination of 5 wt% of GNs results in an increase of several orders of magnitude in the PP/GNs conductive composites, reaching 10^{-6} S/m, which indicates that the conductive composites have the percolation phenomenon; namely, a conductive network is formed between the conductive fillers. According to the percolation theory, the percolation threshold of PP/GNs conductive composites is 2.8 vol%. An increasing content of GNs results in a more complete conductive network and a greater conductivity. When combined with 20 wt% of GNs, the conductivity of PP/GNs conductive composites reaches 0.36 S/m. Figure 9 shows that PP/CF conductive composites containing 3 wt% of CF have a conductivity of 10^{-6} S/m. Moreover, the conductivity constantly retains the same regardless of the increasing content of CF. Such results are likely due to the initial properties of CF.

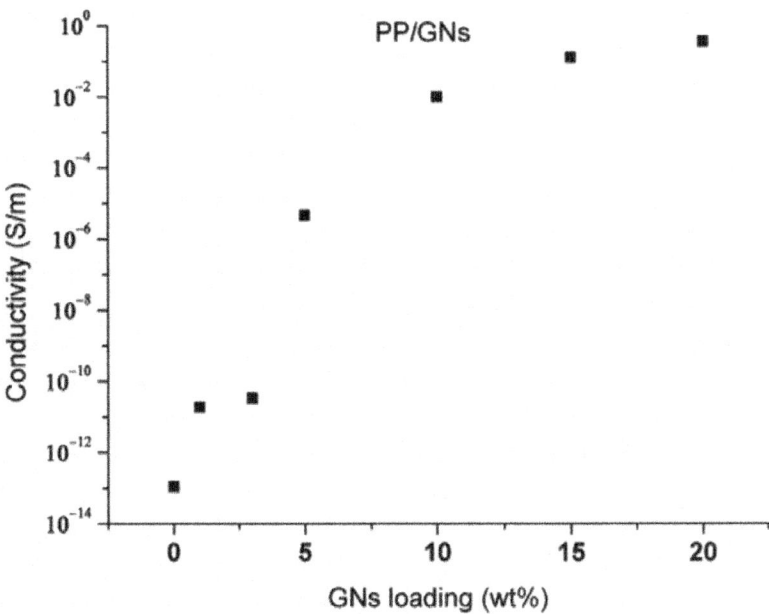

Figure 8. Conductivity of PP/ GNs composites as related to various blending ratios.

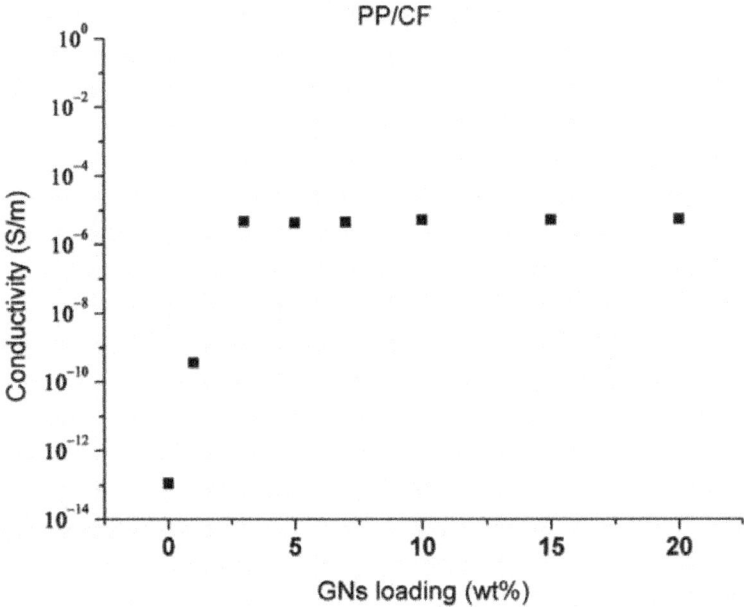

Figure 9: Conductivity of PP/ CF composites as related to various blending ratios.

The EMI SE of polymer conductive composites is correlated with factors such as the electrical conductivity and magnetic conductivity of composites, the thickness and geometry of shield, frequency, the distance between the shield and the field source, and the dispersion of the conductive fillers in composites. Figure 10 and Figure 11 show the EMI SE of the polymer conductive composites containing different contents of GNs or CF, respectively. The EMI SE of the polymer conductive composites increases as a result of an increasing content of GNs or CF. Due to an increase in fillers, the conductivity of the polymer conductive composites increases, which in turn augments their EMI SE [32,33,34,35,36,37,38,39,40,41,42]. At some specified frequencies, the EMI SE appears to be really high, and the extremum value of both PP/GNs and PP/CF conductive composites occur at a similar frequency, which is surmised to be determined by the initial properties of fillers belonging to the carbon category [18,43].

Figure 8 and Figure 9 show that with the same loading level that is more than 5 wt%, the conductivity of PP/GNs conductive composites is greater than that of PP/CF conductive composites. However, the EMI SE of the PP/CF conductive composites is greater than that of the PP/GNs conductive composites as seen in Figure 10 and Figure 11. In sum, the EMI SE of a shield is correlated with not only its conductivity, but also its initial properties.

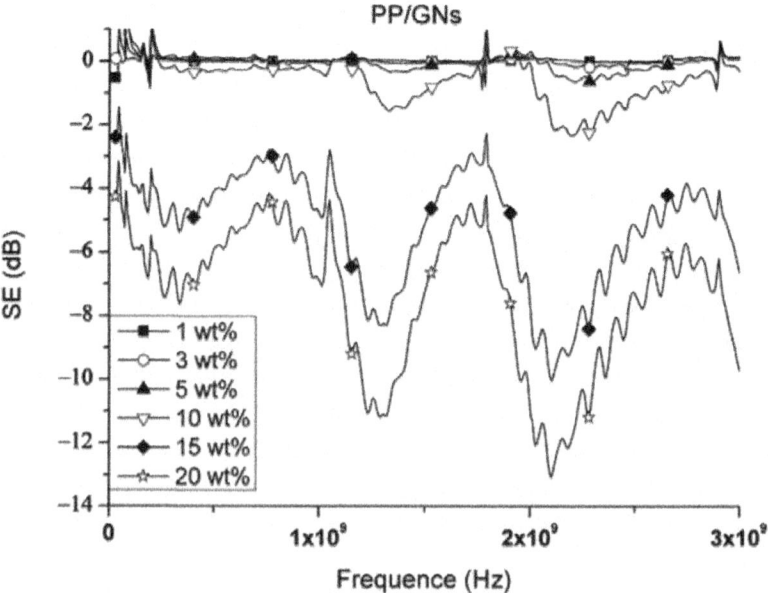

Figure 10: EMI SE of PP/GNs composites with a thickness of 0.5 mm as related to various blending ratios.

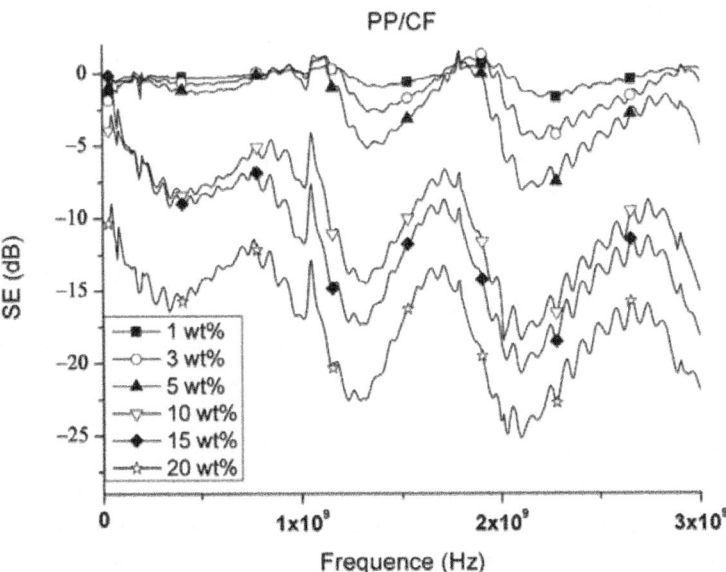

Figure 11: EMI SE of PP/CF composites with a thickness of 0.5 mm as related to various blending ratios.

CONCLUSIONS

This study successfully produces PP/GNs and PP/CF conductive composites by cooperating with a melt compounding method. The combination of these two filler types as the nucleating agent results in a greater crystallization temperature of PP. With the same content, GNs demonstrate a greater influence over the crystallization temperature than CF does. The combination of GNs retards the crystallization rate of PP, while the combination of CF does the opposite. The tensile modulus of both PP/GNs and PP/CF conductive composites is considerably proportional to the loading level of the conductive fillers. The tensile strength of PP/GNs conductive composites decreases with an increasing content of GNs, while that of PP/CF conductive composites first decreases then increases as a result of the increasing CF. The optimal conductivity of 0.36 S/m and the optimal EMI SE of 13 dB occurs when the PP/GNs conductive composites contain 20 wt% of GNs. By contrast, the optimal conductivity of 10^{-6} S/m occurs when the PP/CF conductive composites contain more than 3 wt% of CF, while the optimal EMI SE occurs when containing 20 wt% of CF. Therefore, possible applications in domestic appliances of the composites in terms of electrical and mechanical properties can thus be attained by adjusting different fillers and different amounts of them.

ACKNOWLEDGMENTS

The authors would especially like to thank Ministry of Science and Technology of Taiwan, for financially supporting this research under Contract MOST 103-2221-E-035-027.

AUTHOR CONTRIBUTIONS

In this study, the concepts and designs for the experiment, all required materials, as well as processing and assessment instrument are provided by Jia-Horng Lin and Ching-Wen Lou. Data are analyzed, and experimental results are examined by Chen-Hung Huang, Chien-Lin Huang, and Chi-Fan Liu. The experiment is conducted and the text is composed by Xiao-Min Song.

REFERENCES

1.   Shen, L.; Wang, F.Q.; Yang, H.; Meng, Q.R. The combined effects of carbon black and carbon fiber on the electrical properties of composites based on polyethylene or polyethylene/polypropylene blend. *Polym. Test.* 2011, *30*, 442–448.

2.   Huang, J.; Mao, C.; Zhu, Y.; Jiang, W.; Yang, X. Control of carbon nanotubes at the interface of a co-continuous immiscible polymer blend to fabricate conductive composites with ultralow percolation thresholds. *Carbon* 2014, *73*, 267–274.

3.   Shrivastava, N.K.; Khatua, B.B. Development of electrical conductivity with minimum possible percolation threshold in multi-wall carbon nanotube/polystyrene composites. *Carbon* 2011, *49*, 4571–4579.

4.   Chen, Y.-J.; Dung, N.D.; Li, Y.-A.; Yip, M.-C.; Hsu, W.-K.; Tai, N.-H. Investigation of the electric conductivity and the electromagnetic interference shielding efficiency of SWCNTs/GNS/PAni nanocomposites. *Diamond Relat. Mater.* 2011,*20*, 1183–1187.

5.   Zhao, S.; Zhao, H.; Li, G.; Dai, K.; Zheng, G.; Liu, C.; Shen, C. Synergistic effect of carbon fibers on the conductive properties of a segregated carbon black/polypropylene composite. *Mater. Lett.* 2014, *129*, 72–75.

6.   Calberg, C.; Blacher, S.; Gubbels, F.; Brouers, F.; Deltour, R.; Jerome, R. Electrical and dielectric properties of carbon black filled co-continuous two-phase polymer blends. *J. Phys. D* 1999, *32*, 1517–1525.

7.   Wen, M.; Sun, X.; Su, L.; Shen, J.; Li, J.; Guo, S. The electrical conductivity of carbon nanotube/carbon black/polypropylene composites prepared through multistage stretching extrusion. *Polymer* 2012, *53*, 1602–1610.

8. Drubetski, M.; Siegmann, A.; Narkis, M. Electrical properties of hybrid carbon black/carbon fiber polypropylene composites. *J. Mater. Sci.* 2006, *42*, 1–8.

9. Straat, M.; Boldizar, A.; Rigdahl, M.; Hagstrom, B. Improvement of melt spinning properties and conductivity of immiscible polypropylene/polystyrene blends containing carbon black by addition of styrene-ethylene-butene-styrene block copolymer. *Polym. Eng. Sci.* 2011, *51*, 1165–1169.

10. Pramanik, P.K.; Khastgir, D.; Saharu, T.N. Conductive nitrile rubber composite containing carbon fillers: Studies on mechanical properties and electrical conductive. *Composites* 1992, *23*, 183–191.

11. Thongruang, W.; Spontak, R.J.; Balik, C.M. Correlated electrical conductivity and mechanical property analysis of high-density polyethylene filled with graphite and carbon fiber. *Polymer* 2002, *43*, 2279–2286.

12. Xu, H.-P.; Dang, Z.-M. Electrical property and microstructure analysis of poly(vinylidene fluoride)-based composites with different conducting fillers. *Chem. Phys. Lett.* 2007, *438*, 196–202.

13. Kalaitzidou, K.; Fukushima, H.; Drzal, L.T. A new compounding method for exfoliated graphite–polypropylene nanocomposites with enhanced flexural properties and lower percolation threshold. *Compos. Sci. Technol.* 2007, *67*, 2045–2051.

14. Sengupta, R.; Bhattacharya, M.; Bandyopadhyay, S.; Bhowmick, A.K. A review on the mechanical and electrical properties of graphite and modified graphite reinforced polymer composites. *Prog. Polym. Sci.* 2011, *36*, 638–670.

15. McLachlan, D.S.; Chiteme, C.; Park, C.; Wise, K.E.; Lowther, S.E.; Lillehei, P.T.; Siochi, E.J.; Harrison, J.S. AC and DC percolative conductivity of single wall carbon nanotube polymer composites. *J. Polym. Sci., Part B: Polym. Phys.* 2005, *43*, 3273–3287.

16. Miyazaki, K.; Okazaki, N.; Nakatani, H. Improvement of electrical conductivity with phase-separation in polyolefin/multiwall carbon nanotube/polyethylene oxide composites. *J. Appl. Polym. Sci.* 2013, *128*, 3751–3757.

17. Stankovich, S.; Dikin, D.A.; Dommett, G.H.; Kohlhaas, K.M.; Zimney, E.J.; Stach, E.A.; Piner, R.D.; Nguyen, S.T.; Ruoff, R.S. Graphene-based composite materials. *Nature* 2006, *442*, 282–286. [PubMed]

18. Potts, J.R.; Dreyer, D.R.; Bielawski, C.W.; Ruoff, R.S. Graphene-based polymer nanocomposites. *Polymer* 2011, *52*, 5–25.

19. Geim, A.K.; Novoselov, K.S. The rise of graphene. *Nat. Mater.* 2007, *6*, 183–191. [PubMed]

20. Hsiao, S.-T.; Ma, C.-C.M.; Tien, H.-W.; Liao, W.-H.; Wang, Y.-S.; Li, S.-M.; Huang, Y.-C. Using a non-covalent modification to prepare a high electromagnetic interference shielding performance graphene nanosheet/ water-borne polyurethane composite. *Carbon* 2013, *60*, 57–66.

21. Zhang, H.-B.; Zheng, W.-G.; Yan, Q.; Yang, Y.; Wang, J.-W.; Lu, Z.-H.; Ji, G.-Y.; Yu, Z.-Z. Electrically conductive polyethylene terephthalate/graphene nanocomposites prepared by melt compounding. *Polymer* 2010, *51*, 1191–1196.

22. Lin, Y.; Chen, H.; Chan, C.M.; Wu, J. Nucleating effect of calcium stearate coated $CaCO_3$ nanoparticles on polypropylene. *J. Colloid Interface Sci.* 2011, *354*, 570–576. [PubMed]

23. Sui, G.; Fuqua, M.A.; Ulven, C.A.; Zhong, W.H. A plant fiber reinforced polymer composite prepared by a twin-screw extruder. *Bioresour. Technol.* 2009, *100*, 1246–1251. [PubMed]

24. Gao, Y.; Wang, Y.; Shi, J.; Bai, H.; Song, B. Functionalized multi-walled carbon nanotubes improve nonisothermal crystallization of poly(ethylene terephthalate). *Polym. Test.* 2008, *27*, 179–188.

25. Xu, Y.; Shang, S.; Huang, J. Crystallization behavior of poly(trimethylene terephthalate)-poly(ethylene glycol) segmented copolyesters/multi-walled carbon nanotube nanocomposites. *Polym. Test.* 2010, *29*, 1007–1013.

26. Ke, K.; Wang, Y.; Yang, W.; Xie, B.-H.; Yang, M.-B. Crystallization and reinforcement of poly (vinylidene fluoride) nanocomposites: Role of high molecular weight resin and carbon nanotubes. *Polym. Test.* 2012, *31*, 117–126.

27. Xu, D.; Wang, Z. Role of multi-wall carbon nanotube network in composites to crystallization of isotactic polypropylene matrix. *Polymer* 2008, *49*, 330–338.

28. Razavi-Nouri, M.; Ghorbanzadeh-Ahangari, M.; Fereidoon, A.; Jahanshahi, M. Effect of carbon nanotubes content on crystallization kinetics and morphology of polypropylene. *Polym. Test.* 2009, *28*, 46–52.

29. Zhao, S.; Chen, F.; Huang, Y.; Dong, J.-Y.; Han, C.C. Crystallization behaviors in the isotactic polypropylene/graphene composites. *Polymer* 2014, *55*, 4125–4135.

30. Korayem, A.H.; Barati, M.R.; Simon, G.P.; Zhao, X.L.; Duan, W.H. Reinforcing brittle and ductile epoxy matrices using carbon nanotubes masterbatch. *Composites Part A* 2014, *61*, 126–133.

31. Prashantha, K.; Soulestin, J.; Lacrampe, M.F.; Krawczak, P.; Dupin, G.; Claes, M. Masterbatch-based multi-walled carbon nanotube filled polypropylene nanocomposites: Assessment of rheological and mechanical properties. *Compos. Sci. Technol.* 2009, *69*, 1756–1763.

32. Al-Saleh, M.H.; Sundararaj, U. Electromagnetic interference shielding mechanisms of CNT/polymer composites.*Carbon* 2009, *47*, 1738–1746.

33. Arjmand, M.; Apperley, T.; Okoniewski, M.; Sundararaj, U. Comparative study of electromagnetic interference shielding properties of injection molded *versus* compression molded multi-walled carbon nanotube/polystyrene composites. *Carbon* 2012, *50*, 5126–5134.

34. Arjmand, M.; Mahmoodi, M.; Gelves, G.A.; Park, S.; Sundararaj, U. Electrical and electromagnetic interference shielding properties of flow-induced oriented carbon nanotubes in polycarbonate. *Carbon* 2011, *49*, 3430–3440.

35. Cao, J.-P.; Zhao, J.; Zhao, X.; You, F.; Yu, H.; Hu, G.-H.; Dang, Z.-M. High thermal conductivity and high electrical resistivity of poly(vinylidene fluoride)/polystyrene blends by controlling the localization of hybrid fillers. *Compos. Sci. Technol.* 2013, *89*, 142–148.

36. D'Aloia, A.G.; Marra, F.; Tamburrano, A.; de Bellis, G.; Sarto, M.S. Electromagnetic absorbing properties of graphene–polymer composite shields. *Carbon* 2014, *73*, 175–184.

37. Mahmoodi, M.; Arjmand, M.; Sundararaj, U.; Park, S. The electrical conductivity and electromagnetic interference shielding of injection molded multi-walled carbon nanotube/polystyrene composites. *Carbon* 2012, *50*, 1455–1464.

38. Thomassin, J.-M.; Jerome, C.; Pardoen, T.; Bailly, C.; Huynen, I.; Detrembleur, C. Polymer/carbon based composites as electromagnetic interference (EMI) shielding materials. *Mater. Sci. Eng. R* 2013, *74*, 211–232.

39. Bian, J.; Lin, H.L.; He, F.X.; Wei, X.W.; Chang, I.T.; Sancaktar, E. Fabrication of microwave exfoliated graphite oxide reinforced thermoplastic polyurethane nanocomposites: Effects of filler on morphology, mechanical, thermal and conductive properties. *Composites Part A* 2013, *47*, 72–82.

40. Deetuam, C.; Samthong, C.; Thongyai, S.; Praserthdam, P.; Somwangthanaroj, A. Synthesis of well dispersed graphene in conjugated poly(3,4-ethylenedioxythiophene):polystyrene sulfonate via click chemistry. *Compos. Sci. Technol.* 2014,*93*, 1–8.

41. Garzon, C.; Palza, H. Electrical behavior of polypropylene composites melt mixed with carbon-based particles: Effect of the kind of particle and annealing process. *Compos. Sci. Technol.* 2014, *99*, 117–123.

42. Noel, A.; Faucheu, J.; Rieu, M.; Viricelle, J.-P.; Bourgeat-Lami, E. Tunable architecture for flexible and highly conductive graphene–polymer composites. *Compos. Sci. Technol.* 2014, *95*, 82–88.

43. Zhao, S.; Chen, F.; Zhao, C.; Huang, Y.; Dong, J.-Y.; Han, C.C. Interpenetrating network formation in isotactic polypropylene/graphene composites. *Polymer* 2013, *54*, 3680–3690.

Chapter 10

NEW SUPERHARD TERNARY BORIDES IN COMPOSITE MATERIALS

Zachary Zachariev
Institute of Polymers, Bulgarian Academy of Sciences, Bulgaria

INTRODUCTION

Superhard substances are those a hardness above 20 GPa, i.e. higher than that of corundum (Kislii et al.,1988) or with a Vickers hardness Hv exceeding 40 GPa (Sologenko&Gregorianz, 2005). The ten utmost non-metal substances and refractory compounds form a "hardness pyramid" (Kislii et al.,1988) diamond being on top, followed by cubic-boron nitride and boron carbide.

Metal-like ones form a similar pyramid, the transition metal borides occupying its top. However, the maximum hardness found for them is inferior to that of the non-metal substances. The interest in borides is due to their extraordinary hardness (up to 1873 °C) as compared to other refractory compounds.

The hardest boride ($B_{12}C_3$) is used as a wear resistant polycrystalline material, armor tiles, nuclear industry, etc. (Anderson, 2002).However, applications are restricted by its high brittleness due to the strong covalent bonds in its crystal lattice. It has been established (Zakhariev&Radev, 1988) that superhard ternary compounds based on boron carbide with dissolved IV-VI group metals ($B_{12-n}C_3Me_n$) can be obtained by sintering CMC composite materials without pressure.

Sintering of TiB_2-MeC-Me systems with no pressure applied is a new trend in the field of superhard boride composite materials. The binding metal (e.g. Ni or Co) in TiB_2-Me system reacts with the boride and forms a low-melting phase (Lecrivan&Provost, 1968); (Fitzer, 1973). The growing interest in borides comes from their high-temperature behavior: high melting point, hardness, wear resistance and chemical inertness. These allows using them to produce cathodes for electrolytic aluminium, first wall coatings and neutron absorbers for nuclear technology, valve components in cool liquefaction plants and crucible materials for metal evaporation.

Studies on the TiB_2-TiC and TiB_2-TaC systems show that the carbides inhibits the grain growth of the boride phase (Murata et al., 1967).The TiB_2-MeC-Me materials have the following advantages as compared with the binary systems mentioned (Petzow&Telle, 1984): application of hot pressing is not necessary, the grain growth is completely inhibited, the mechanical properties are improved due to the small grain size (about 1μm) (Zakhariev et al., 1993).

WC-Co is the source material in the field of the metal-working whereas superhard ternary boride (WCoB) coating entails a sharp rise in wear resistance of the composite materials (Zakhariev et al., 1987). The micro hardness of synthetic polycrystalline WCoB amounts to 4650 кgf/mm² (Zakhariev et al., 1986) which explains the increased service tool - life of the WC-Co instruments. Similar ternary compounds are MoCoB and WFeB with the same orthorhombic unit cells (Jeitschko, 1968).

Combining borides (B_4C, TiB_2 et al) with carbides (MeC, WC-Co et al) allows their chemical interactions upon heating, the building up of eutectics etc. This can result in densification of the composite material without applying of any additional pressure as well as in enhanced physico-mechanical properties due to the arising of superhard ternary compounds ($B_{12-n}C_3Me_n$, $(Ti,W)_2B$, WCoB) (Zachariev, 2001).

Under thermochemical treatment of steels (Fe-C) with a transition metal borides (ZrB_2, TiB_2, CrB2) (Zakhariev et all., 1970) it appears a ternary compound $(Fe,Me)_2B$ of considerable hardness 20-23 GPa. The layer it has built up combines the advantages of consecutive metalizing and boronizing layers thus bringing about the enhanced resistance to wear and corrosion of the metal-matrix composite (MMC) system.

Similar properties with respect to corrosion and wear have another complex boride (synthesized Mo_2FeB_2 with ferrous binder), known as "new hard alloy" (Komai et al., 1989). The ternary boride $Cr_xFe_yB_z$ ($Cr_3Fe_{80}B_{17}$) also shows unique properties which have been utilized in several amorphous materials for corrosion applications.

Fukatsu et al. (Fukatsu, 1967) have shown that the wear resistance of hard alloys increases with the increasing of their hardness provided all other conditions are the same.

The present paper, based mainly on research done by the author and associated colleagues, aims at a concise unification of the results on new superhard boride composites in view of their practical applications.

EXPERIMENTAL METHODS

The carbides ($B_{12}C_3$, TiB_2 and $Me^{IV-VI}C$ produced by "ESK-Kempten", "Merck" and "Ventron Alfa Products", respectively) were homogenized in a Frisch planetary mill, pressed at 200 MPa and then sintered at 1700 - 2250°C in a Degussa furnace with graphite heater in argon atmosphere.

The hardness was determined using a "Leitz-Durimet" hardness tester with loads of $HV_{0.5}$ to HV_1 and a "Carl Zeiss" micrometer with indentation load 30-100 g. Compound hardness as a function of indentation load has been traced.

An automated DRON-3 diffractometer, with CuKα radiation was used to investigate the structure of the materials under study. Their morphology was characterized by scanning electron spectroscopy (SEM) on a JSM 840 apparatus equipped with a Link QK 200 dispersive X-ray analyzer.

The initial TiB_2 powder (type PIII; Ti-68.6, B-27.3, TiO_2-3.9, Co-0.2 wt.%) was prepared under industrial conditions using a technology developed in the Institute of problems in Materials Science, Kiev. The chemical analysis of another initial TiB_2 Koch-Light powder showed Ti-68.8 and B-31.2 wt.%. The powders TiB_2 were milled 75-120 min using hard-alloys bodies and vessels of WC-Co (K10) in the Fritsch planetary mill with acetone serving as a medium. The polycrystalline samples of the ternary boride WCoB were obtained by crystallization from a cobalt-rich melts of the corresponding powdery components at 1600 °C according to (Petrov&Will, 1981).

Standard WC-Co cutting plates K10 (92 wt.%WC, 6%Co and 2 wt.% TaNbC) were packed separately in Borozar-HM (powdery product, 325 mesh) and B_4C (F220 technical grade, ESK-Kempten) and heated at 1000-1400°C for 30-120 min in an inert medium (argon). The heating was carried out in a large-scale Bor 6-CM-3 installation for deposition of boride coating on K10 plates.

For simultaneous boron-zirkonizing (or boron-chromizing) of steel samples by powdery borides $ZrB_2(CrB_2)$ and some activators were used the heating occurring at 950-1050°C in argon atmosphere.

Obtaining the superhard boron-metallizing layer on steel tools or parts of them requires their coating in a patented paste (commercial paste "Zahobor-P"). The technology is very simple: a coat of the paste is applied over the working surfaces of machine tools and parts. The metal surfaces thus coated are dried and then subjected to heat-treatment. Additional procedures, staff and equipment are not required.

RESULTS AND DISCUSSION

Ternary $B_{12-N}C_3ME_N$ Borides in CMC-Composite $B_{12}C_3 + ME_xB_Y$

It is obvious that a sintering of $B_{12}C_3$ is only possible at temperatures above 2100°C, i.e. close to the melting point of boron carbide (2447°C). SEM of a fracture surface of $B_{12}C_3$ (10 wt %) + W_2B_5 material sintered at 2150°C and a grain size of 2-7 μm is shown in Fig.1. With larger magnifications (2000x), a thin eutectic binder is visible at the grain boundaries.

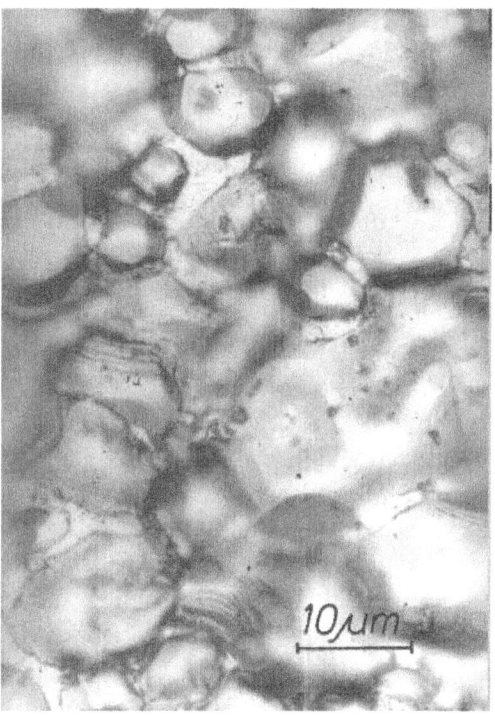

Figure 1: SEM of the fracture surface of $B_{12}C_3$ + 10 wt%W_2B_5 composite material (sintered at 2150°C; 20 min).

An eutectic resulting from interaction between the two carbides (B_4C + WC → W_2B_5 + C) does not seem unexpected when taking into account the eutectic character of the ternary diagrams B-C-Me [IV-VI] (Schouler et al.,1983) and the quasi-binary systems B_4C-MeB$_2$ (Portnoi& Samsonov,1960).

Data in Tabl.1 show that the systems containing 4th Period metals, such as $B_{12}C_3$-TiB$_2$ (VB$_2$,CrB$_2$) have the lowest T_{eut} (2150-2200 °C) as compared with the other ones ($B_{12}C_3$-ZrB$_2$ /NbB$_2$ and $B_{12}C_3$-HfB$_2$/ TaC), which is of importance for the assessment of the sintering temperature of composites.

Table 1: Melting point and Eutectic temperature of composite system $B_{12}C_3$-MeB_2

MeB2 in eutectic, %				
Compounds Tm,oC	ΔH,KJ/mol	vol.	Mol	Teut, oC
B12C3	2447	70,0	-	-	-
TiB2	3217	280,0	20,0	26	2197
ZrB2	3247	314,0	20,0	24	2277
HfB2	3347	335,0	23,0	22	2377
VB2	2747	142,4	35,0	46	2167
NbB2	2997	174,6	35,6	36	2247
TaB2	3097	217,7	27,0	32	2367
CrB2	2217	125,6	63,0	70	2147

Due to the Me_xB_y particles it creates, appearance of the eutectic leads to brittle boron carbide becoming stronger. This results in a significant decrease in brittleness of the recrystallized composite and an increase in its crack-resistance. Measured by the indentation method, the critical coefficient of stress intensity (KIc) is in differently oriented eutectics ($B_{12}C_3$-MeB_2) ranges from 6-12 MPa to 2.5-3.5 MPa depending on their components. This result indicates that the use of eutectics as a new class of CMC composite materials resistant under extreme conditions is promising.

The microhardness values of a ternary phase $B_{12-n}C_3Me_n$ borides in composite $B_{12}C_3 + Me_xB_y$, as compared to hot-pressed pure boron carbide, and the values obtained by other authors are presented in Table 2. The data show that the boron carbide hardness in the ternary borides phases is much higher than that of pure boron carbide and other composites based on $B_{12}C_3$, c-BN and c-BC_2N.

This sharp increase in hardness seems to be due to dissolution of the transition metals in the crystal lattice of boron carbide. During the chemical interaction of the two carbides the boron needed for the reaction comes from the boron carbide lattice.

Since from the investigations of Lipp et al. (Lipp&Schwetz, 1975) it is known that the homogeneity region of $B_{4+x}C$ has no substantial effect on its hardness, the only reason for the sharp increase in boron carbide hardness should be the formation of new superhard ternary compounds with the formula $B_{12-n}C_3Me_n$ (Fig. 2).

These compounds do not differ essentially from the ternary compounds $B_{12}C_{3-n}Me_n$ (predicted byLipp &Roder, 1966 and proved to Okada et al.,

1990). The difference is that with the latter compounds, the replacement of Co (at. Radius 0.91 Å) by the larger element Al (at. radius 1.43 Å) leads to a considerable increase in the size of the unit cell in the direction of the hexagonal C- axis and of the lattice volume (Table 3). In our case, just the opposite happens. The volume shows no substantial change, whereas the lattice parameter c decreases (see Table 3).

This is attributed to an increase in carbon content in the homogeneity region of the boron carbide, which is due to elimination of boron from the compound. The decrease in the parameter co, while the crystal lattice volume remains unchanged, indicates that the metal atoms are incorporated most probably in the icosahedron B2 and B1 sites.

Table 2: Microhardness (GPa) of $B_{12-n}C_3 Me_n$ from the initial materials $B_{12}C_3 + Me_{x-}$ B_y and other composites

Initial Material (weight %)	HV1	HV05	Reference
B12C3+W2B5 (10 ; 50)	50; 58	53* ; 77**	(Zakhariev, 1988)
B12C3+CrB2 (10)	56	77**	(Zachariev, 2001)
B12C3 + VB2 (10)	43	52	(Zachariev, 2001)
B12C3+Ti (7)	-	63	(Makarenko et al. ,1977)
B12C3+Zr (10)	-	60	(Makarenko et al.,1977)
B12C3+ Al (50)	25	-	(Lipp & Roder ,1966)
B48C2Al3 - crystal	30.5	-	(Okada et al., 1990)
B12C3 + TiB2 (10;20)	-	45 ; 55	(Nishiyama , 1985)
B12C3+TiB2 (10;25)	-	30 ; 34	(Telle & Petzow ,1987)
B12C3+TiB2 (10)+W2B5 (5)	-	37 ; 42	(Telle & Petzow ,1987)
c-BC2N	-	76	(Solozhenko , 2001)
B12C3-HP "pure"	34	30 ; 38 ; 30	(Lipp &Schwetz , 1975)

Length of pyramid diagonal indentation [m]: * 4.4; ** 3.15

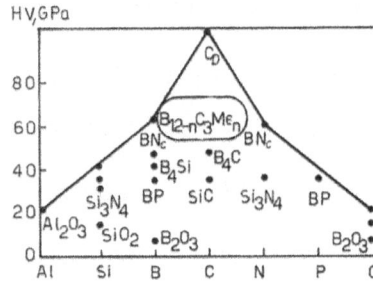

Figure 2: Non-metal hardness pyramid with the new ternary boride phase $B_{12-n}C_3Me_n$.

Table 3: Parameters of the crystal lattice of the ternary metal borides compounds C_0[Å], a_0[Å]; V[nm³]

Compounds	Co[Å]	ao[Å]	Volume [nm3]	Reference
B11C3W (10%W2B5)	12,078	5,601	0,32757	(Zachariev , 2001)
B11C3W (20%W2B5)	12,024	5.604	0,32697	(Zachariev , 2001)
B11C3W (50%W2B5)	12,054	5,606	0,32802	(Zachariev , 2001)
B11C3Ti (10 – 20%)	12,06-12,07	5,607	-	(Zachariev , 2001)
B11C3V (10- 20%)	12,02-12,05	5,605	-	(Zachariev , 2001)
B12C2Al	12,39	5,65	0,34252	(Lipp & Roder , 1966)
B12(C,Si)3	12,1-12,4	5,60	0,3286-0,3367	(Kislii at all. , 1988)
B13C2	12,19	5,67	0,33938	(Thevenot & Bouchacour, 1979)
B12C3 (initial ESK)	12,12	5,60	0,32915	(Lipp & Schvetz , 1975)

Due to the strongly extended lattice during the incorporation of Al and Si into the C-C-C axis, the phases $B_{12}C_2Al$, $B_{48}C_2Al_3$ and $B_{12}C_2Si$ have a microhardness (HV1 = 25 – 30.5 GPa) even lower that that of pure boron carbide (Table 2).

A novel superhard phase c-BC_2N was synthesized using the laser-heated diamond anvil cell with a hardness Hv=76 GPa [Solozhenko, 2001] (Tabl.2).

The hardness (50-77GPa) of the new ternary borides is much higher than that of "pure" boron carbide sintered by hot pressing (Table 2). It is equal to that of cubic boron nitride (i.e. next to diamond) and even of some polycrystalline diamonds of the type "Carbonado" and "CB". However, the price of the new boride is several orders of magnitude lower.

We have developed an original method to obtain articles of boron carbide composite, in which the hot-pressing stage is avoided. The new superhard boron carbide has already found several applications: in the production of armor plates to protect people and machines from bullets and shrapnel (Fig. 3 and Fig. 4), for protection against neutron radiation in nuclear power plants as well as in nuclear therapy of tumours.

Other possible applications will use the large capacity for neutron absorption by the investigated CMC- composites. The cross-sections of neutron absorption are as follows: boron carbide – $1,98.10^{-2}$barn, tungsten boride (W_2B_5) – $2,14.10^{-1}$ barn, the proposed CMC material $B_4C + W_2B_5$ – $3,18.10^{-1}$barn. The latter has been successfully tested in the nuclear power plant in Kozloduy, Bulgaria for 14 months.

Ternary (TI,W)B$_2$ Boride in CMC-Composite Materials TIB$_2$-WC-CO

The milled TiB$_2$-WC powders containing up to 1 wt. % cobalt are sintered in order to obtain two-phase cermets, i.e. to avoid crystallization of the brittle ternary phases WCoB and W$_2$CoB$_2$. The ternary phases were found in sintered samples which are obtained from initial mixtures containing 2 to 10 wt. % Co. The densification of the alloys during the sintering began above 1100 °C. Increasing the temperature, the decomposition of WC causes formation of facets on the grain surfaces. During the isothermal sintering above 1700 °C to 1850 °C, β-WB and (Ti,W)B$_2$ solid solutions were precipitated on the TiB$_2$ nuclei.

The microstructure of the initial powder TiB$_2$ +TiO$_2$+ 18.64WC + 1Co is characterized by homogeneous distribution of fine TiB$_2$ grains (4.4 μm). The presence of the (Ti,W)B$_2$ -phase seems to be more pronounced now as the latter is not restricted to the periphery of TiB$_2$ (the Koch-Light sample) entire grains of it being observed (Fig. 5). Some of the boride crystals have partially lost their boron through the eutectic Co-WCoB film (TiB$_2$: a=3.018; c=3.209Å). The eutectic built up seems uniformly distributed remaining all but invisible with a thickness of less than 0.4 μm. The size of the pores is 1-3 μm.

Figure 3: Protective ceramic equipment.

Figure 4: Composite boroncarbide ceramic plates on armored vehicle (Company Rafael - Israel).

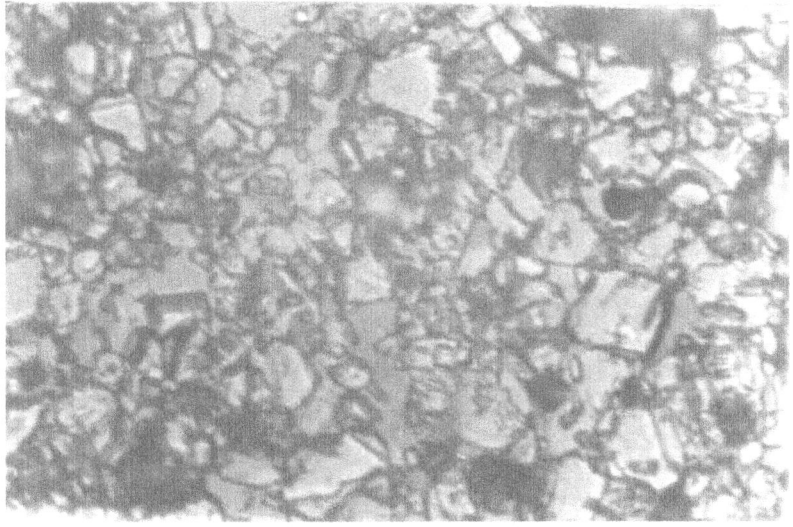

Figure 5: The microstructure of TiB_2- WB – $(Ti,W)B_2$ composite; x 1600.

The mechanical properties of the TiB_2-WB-Co composite materials under study have exhibited a very high hardness combined with a high strength.

Small amounts of the binding metal (cobalt – 1%) considerably decrease the sintering temperature. This is due to reactions which yield eutectics. The same binding metal allows densification of the alloys during sintering without pressure application.

As a general rule, introducing small amounts of the binding metal (1 wt-% with colalt) in TiB_2-WB system considerably decreases the sintering temperature considerably without any resort to pressure while yielding a very high hardness 50.6 GPa (92.3 HRA) (Fig.6 and Tabl.4) of the ternary phase $(Ti,W)B_2$.

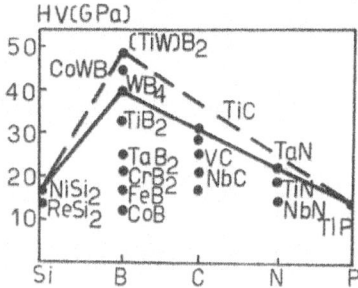

Figure 6: Metal-like hardness pyramid with the new superhard ternary phases CoWB and $(Ti,W)B_2$

The difference in hardness between of the two composite materials (Table 4) seems to come from more ternary $(Ti,W)B_2$ being present in the initial powder TiB_2 (PIII), due to contamination of the latter with TiO_2.

Table 4: Properties of TiB_2-WB-Co sintered in Ar at 1850 °C, 120 min

Initial milled Specific surface Experimental Relative Hardness Micro-hardness					
powder[wt%] area [m2/g] density[g/cm3] density[%] RockwellA Hμ0.1 [GPa]					
TiB2(Koch-Light)	2.58	5.22	98.30	91.3	33.83
+ 20.25WC+0.75Co					
TiB2(PIII)+18.64WC	3.53	5.09	98.80	92.3	50.6
+ 1 Co+ 3.9TiO2					

Ternary WCOB Boride in MMC-Composite Materials WC(TIC)-CO

For the sintered ternary polycrystalline tungsten-cobalt boride WCoB chemical analysis gave: W- 66.3±6 + Co- 24.7±2 + B- 6±3 wt.% while X-ray studies showed that it is orthorhombic WCoB (ordered $PbCl_2$ structure, Type E-TiNiSi), a=5.724Å; b=3.240Å; c=6.632Å.

Several large WCoB crystals (dimensions of 1x1.5x3 mm) are shown in Fig.7-1.

The dependence of the compound hardness on the indentation load was plotted. In this way, the straight line of Meyer and the line of microhardness (Fig.7-2) were obtained. Obviously, within the range of indentor loads used (below 100 gf), the microhardness gradually decreased with increasing loading. Hence, the microhardness value for WCoB depends on the loading. This correlation is due mainly to plastic deformation, which was observed in our case at a low indentation load (less than 50 gf).

The {001} and {100} faces of WCoB showed a reticular anisotropy of microhardness in directions c and b.

The hardness of the ternary compound, $H\mu50 = 4650 \pm 230$ kgf.mm^{-2}, is associated mainly with the type and the distribution of bonds in it and corresponds to the usual high hardness of the transition metal borides. Gilman (Gilman, 1970) is of the opinion that this hardness is due for the most part to overlapping of the metal-nonmetal bonds during the shearing of the dislocations. Within the framework of this model $H\mu=-2\Delta H_f/V_m$, where ΔH_f denotes the heat of formation of borides (kcal mol^{-1}) and V_m is the molecular volume (cm^3). Thereupon, after converting 4650 kgf.mm^{-2} into 5.45 kcal.cm^{-3} (the value of $H\mu$) and 30.758 Å^3 into 18.5 cm^3.mol^{-1} (the value of V_m), one obtains for WCoB $\Delta H_f = 100.5$ kcal mol^{-1}. Comparison of the heat of formation of WCoB with that of TiB_2 (77.4 kcal mol^{-1}) confirms the increase in hardness of the compound with rising heat of formation.

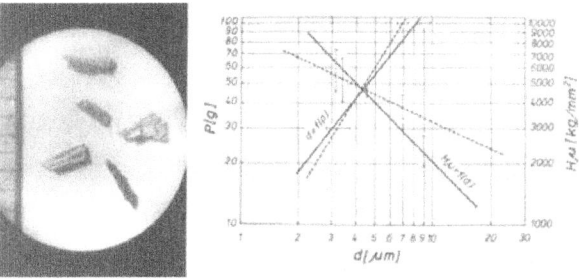

Figure 7: Ternary polycristals WcoB, Figure 7-2. Dependence of Hμ on the indentor loading for WCoB

Formation of WCoB upon the WC(TiC)-Co matrix during thermo-chemical treatment could result from interaction of the type WC + TiB$_2$ + Co → WCoB + TiC + CoB.

TiB$_2$ most probably participates in the coating formation as a donor of boron which diffuses into the samples and interacts with the cobalt and the tungsten carbide. A similar mechanism is proposed in order to explain the formation of boride coatings on iron and steel during thermo-chemical treatment with other boronizing agents (boron and B$_4$C). The concentration of WCoB in the diffusion layer depends on the composition of the initial alloys and the experimental conditions of thermal treatment 950-1100 °C (Fig.8).

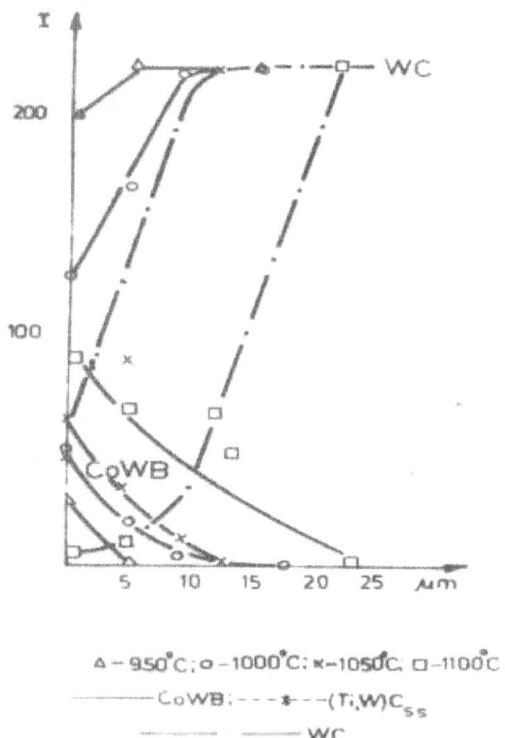

Figure 8: The phase composition us depth through the diffusion layer formed on a WC-Co: Δ- 950 °C; O- 1000 °C; x- 1050 °C; □- 1100 °C; — WCoB; -.- WC.

The presence of the ternary orthorhombic compound WCoB in the surface layer of carbide alloys enhances their wear resistance in metal-cutting. The enhancement seems to follow from the increase in their hardness (Fig. 9. The difference in phase composition of the diffusion layers obtained using the two powders affects the layer hardness. The use of Borozar-HM (base TiB_2) leads to the formation of WCoB only in the diffusion layer, whereas in the case of B_4C the ternary boride W_2CoB_2, which is richer in boron, prevails. This can be explained on the basis of the Co-W-B phase diagram (Stadelmaier, 1967) taking into account the boron content (i.e. the transfer of boron from boron carbide). The termo-chemical treatment with Borozar-HM of the cutting alloy results in the formation on its surface of superhard layers whose hardness exceeds that of the layers obtained with B_4C. The maximum hardness value 23.4 GPa was found for layer with Borozar-HM at 1200 °C, which is assigned to the formation at this temperature of a single-phase ternary WCoB layer.

Ternary borides are useful in drawing plain wire and metalworking where a superhard layer of them is formed by diffusion on the main material consisting of carbide – cobalt alloys WC (TiC) – Co.

Layers of this kind improve the performance of nozzles, turning-lathes and other devices used in drawing and cutting of metal articles such as wire, rods, pipes, plates (Fig.10).

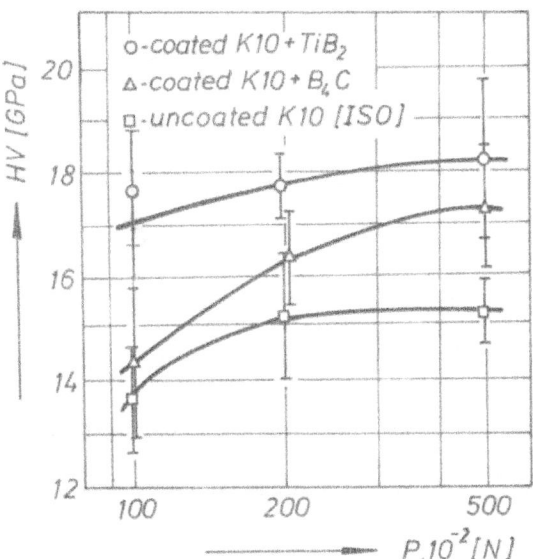

Figure 9: Dependence of the boride layer hardness on the temperature of matrix (K10, WC-Co-10%) treatment with Borozar-HM (TiB_2) and B_4C.

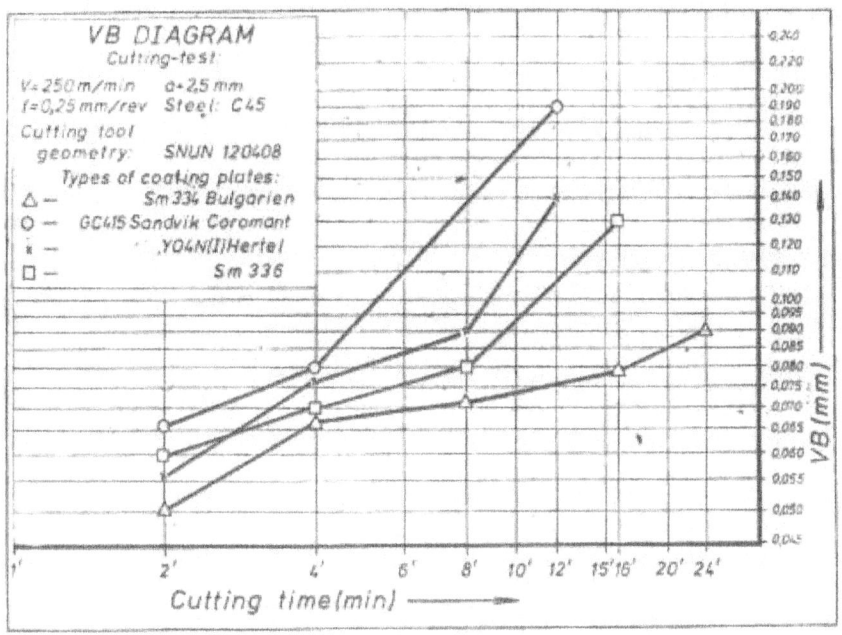

Figure 10: Comparative diagram (VB) of the wear in steel working of various types of coated cutting tools: □ – WCoB+ TiC-Al$_2$O$_3$ deposited on Sm 336 plate (Gabrovo,Bulgaria); X- TiC-Al$_2$O$_3$ deposited on Cp 1 –V04 (Hertel, Germany); O-TiC-Al$_2$O$_3$-TiN deposited on GC 415 plate (Coromant, Sweden) ; Δ- WCoB + TiC-Al$_2$O$_3$ on Sm334 plate (Gabrovo, Bulgaria).

Ternary (FE,ZR)$_2$B Boride in MMC- Composite Material (FE-C Matrix)

The thermochemical treatment with ZrB$_2$ (CrB$_{2,}$ TiB$_2$) and activators on steels yields a diffusion layer with a thickness between 50 μm and 1 mm. Our investigation of ZrB$_2$ (KBF$_4$ having been used as a activator) were carried out over the temperature range 1000-1100 °C for 4 h. The diffusion layer obtained at 1050 °C was of a two-zone nature: zone "A" and the underlying (in the direction of the sample centre) zone "B" (Fig.11(A)and Fig.11 (B), respectively). The photograph of the polished section in Fig.11(A) shows that zone "b" consist of needle-like crystals characteristic for the gas transport of boron. Zone "B", which is situated above layer b-d, has a different structure with small grain size and a very high microhardness (3575-2438 kg.mm^{-2} (Fig. 11(B)).

For the zones below the zirconium a greater microhardness (as compared with that of "pure" Fe$_2$B) has been detected by microprobe analysis. It might be due to replacement of some iron atoms of the Fe$_2$B-phase by zirconium ones. This is also indicated by the change in lattice parameters of the underlying iron boride.

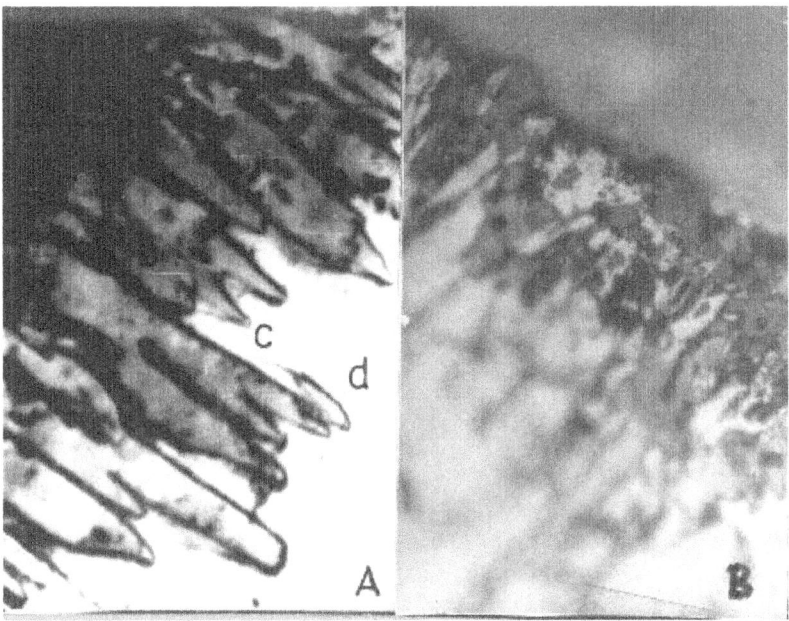

Figure 11: The two-zone diffusion layer on Armco-iron treated with ZrB_2 and KBF_4(3wt.%): A - boronized zone (Fe_2B) ; B - upper superhard $(Fe,Zr)_2B$ - ZrO_2 zone; x 600.

The ternary boride $(Fe,Zr)_2B$ enhances the wear- and heat resistance of the steels coated.

With boron-chromizing, the ternary boride $(Fe,Cr)_2B$ imparts additionally augmented resistance to corrosion (Fig.12).

The microstructure of the boron-chromium layer obtained on steel C45 with Zahobor paste is presented in Fig.13. X-ray microanalysis has shown that black grains of chromium- iron boride $(Fe,Cr)_2B$ contain 14-50 wt.%Cr.

Boron-metalizing with paste during hardening of steel is a new process resulting in a surface layer with a high wear resistance and stability with respect to oxidation and corrosion, minimum time and cost losses needed. The proces carried out with pastes for boronization and boronmetalization leads to products of higher hardness, i.e. durability (twice as long as in cases of nitration and cementation), a higher stability towards high temperatures and a higher corrosion resistance.

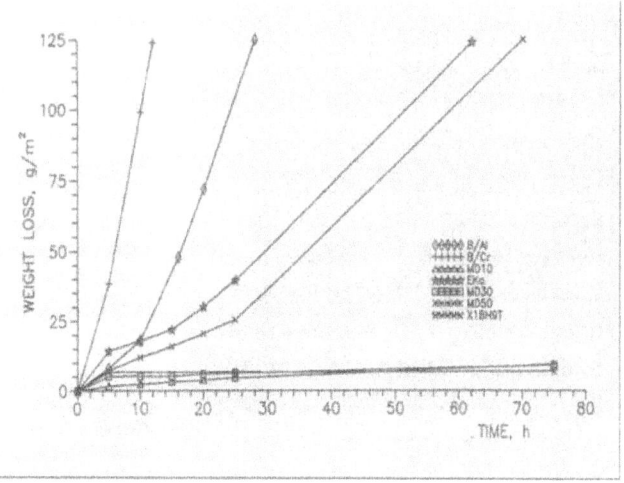

Figure 12: Corrosion resistance in $10\%H_2SO_4$ of boron-chromium (Zahobor paste MD10, MD30, MD50); boronizing with Ekabor-Paste (Eka-Germany); subsequently boronized electrolyte chromium (B/Cr); boro-aluminizing (B/Al) layers deposited on C45 and chrom-nickel steel (X18H9T).

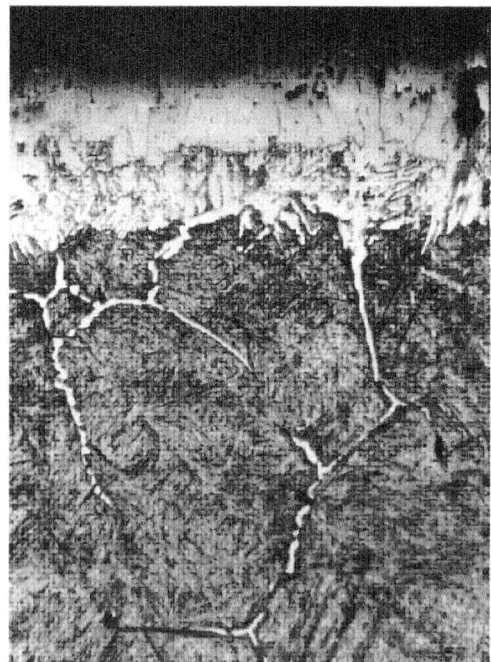

Figure 13: Microstructure of boron-chromium layer deposited on C45 steel by Zahobor paste (MD30); X600.

Table 5 shows the results of x-ray phase analysis of the layers as well as data on their thickness and microhardness. Obviously, the highest hardness corresponds to borchromium layer obtained using Zahobor (MD30).

Table 5: X-ray phase analysis, thickness and microhardness of the diffusion layers with pastes on C45

Paste	Phase composition	Thickness, μm	Hμ 30, GPa	Hμ50, GPa
Zahobor –Bulgaria	CrB, (Fe,Cr)2B	150	20.4	18.6
Ekabor-Germany	Fe2B	160	15.6	15

On the basis of these results it may be inferred that doping of the layers with chromium, which leads to the appearance of $(Fe,Cr)_2B$ phase increases significantly their microhardness. The phase composition of the diffusion layers determines their microhardness, i.e. their wear resistance. Hence, we may predict an even higher wear resistance of our boron-chromium layers. Indeed, the results on their wear resistance correlate with those on their microhardness (Fig.14). The most stable Zahobor (MD30) layer is more than twice as stable as the boronized one according to the fifty-hours-test.

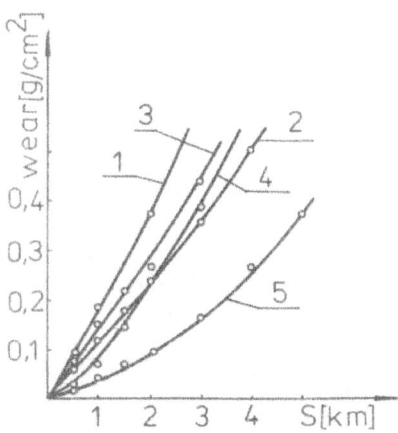

Figure 14: Wear resistance of the boron-chrominized (5-MD30), boronized (4-Ekabor), boron-aluminized (2- B/Al), electrodeposited chromium (3) layers and base chromium-nickel steel.

The positive effect from boron-chromizing is illustrated on Fig.15 and Fig.16 for landholder's steel instruments treated with the "Zahobor"-paste and used in the Netherlands.

The paste is suitable for treatment of steel machine tools and parts of large dimensions, e.g. metal stamps, hammering press matrices, guides, rolls for wiredrawing, steel pulleys, steel belt conveyor rolls, ploughshares, tracks, extruder screws and other similar machine parts, subjected to wear and corrosion. Machine tools, instruments and parts with larger design tolerances as regards cross-section dimensions are especially suitable for boron-metalizing.

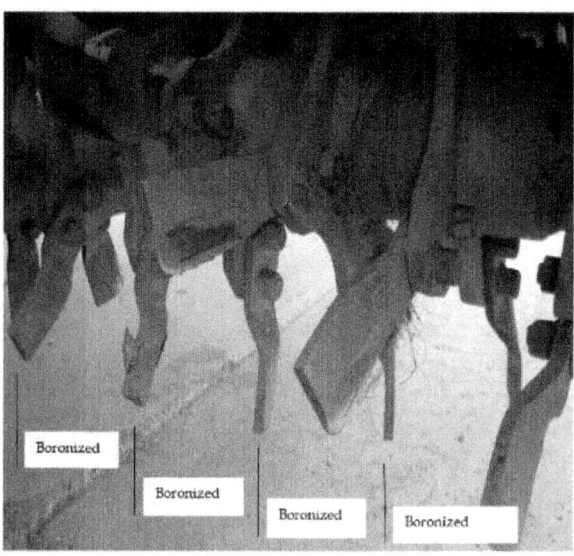

Figure 15: Boronized parts have been used for 6 ha.

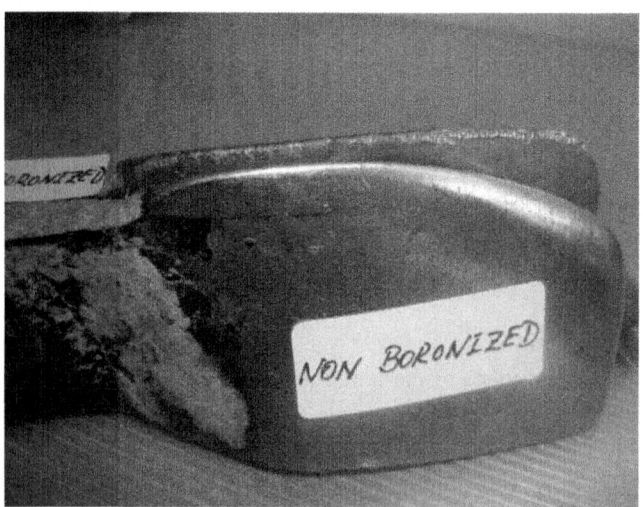

Figure 16: The knifes "Tortella" - Italy boronized with paste "ZAHOBOR".

CONCLUSION

Sintering without applying a pressure is a new trend in the field of superhard boride materials. The present paper deals with the microhardness of some boride CMC and MMC composite materials ($B_{12}C_3$ + Me_xB_y, TiB_2 + WB + Co, WCoB + WC–Co, $(Fe,Me)_2B$ + Fe-C) obtained in this way.

It is shown that the transition metal dissolves in the crystal lattice of $B_{12}C_3$ building up new superhard ternary borides $B_{12-n}C_3Me_n$ with a hardness of 50 - 77 GPa. The latter values exceed considerably the hardness of pure $B_{12}C_3$ and coincide with those for cubic-BN and some synthetic diamonds of the type "CB" or "Carbonado ACPK".

Another example is the composite material TiB_2 - WB, where the surface of its grains proves enriched of tungsten to $(Ti,W)B_2$, this leading to an extremely high value of 50.6 GPa.

The hardness of the ternary boride WCoB amounts to 38 - 43 GPa depending on the indentor loading. Presipitate in the form of a boronizing coating upon the carbide cutting tools, WCoB- phase increase their tool-life.

Thermochemical treatment of steels with ZrB_2 (TiB_2 or CrB_2) leeds to form a diffusion layer with superhard ternary phases $(Fe,Me)_2B$. This phases improve the wear- and corrosion resistance of the steels.

In comparison with other metal- like refractory compounds, the superhardness of the materials studied points to new applications in industry. In this field of view, include my own researches, is make an attempt to unification of the scientific results and to show the perspectives about using of the obtaining superhard ternary composite (CMC and MMC) materials.

REFERENCES

1. Ch. Anderson, 2002 Ceramic armor materials by design, The Amer. Ceram.Soc.,487489

2. E. Fitzer, 1973 Arch.Eisenhuttenwes, 44 703709

3. T. Fukatsu, K. Yuhara, K. Kobori, 1967 Nippon Kinzoku Gakkhaishi, N3 11271131

4. J. Gilman, 1970 J.Appl.Phys., N41, 16641669

5. W. Jeitschko, 1968 Acta Cryst., B24, 930934

6. P. Kislii, M. Kuzenkova, N. Bondarchuk, 1988 Carbide bora, Naukova dumka, Kiev

7. M. Komai, et al. 1989 MRS Int'lMtg.on Adv.Mats., 4 Materials Research Society, 475480

8.   L. Lecrivian, G. Provost, 1968 Berichte der Deutschen Keramischen Gesellschaft, 45 7, 347351

9.   A. Lipp, K. Schwetz, 1975 Berichte Dt. Keram. Ges, N52, 335340

10. A. Lipp, M. Roder, 1966 Z.Anorg. Algemeine Chem., 343 19

11. G. Makarenko, T. Kosolapova, E. Marek, 1977 Tugoplavkie boridi I silizidi AN USSR, Naukova Dumka, Kiev, 6677

12. Y. Murata, H. Julien, E. Whitney, 1967 Ceramic Bulleting 46, N7, 643648

13. K. Nishiyama, 1985 JSCM N11, 5361

14. Sh. Okada, K. Kudou, H. Hiyoshi, I. Higashi, T. Lundstrom, 1990 J.of the Ceram.Society of Japan, Int.Edition, 98-1342 , 4247

15. K. Petrov, G. Will, 1981 J.Materials Science, 16 32183223

16. G. Petzow, R. Telle, 1984 New Development in the Field of Refractory Hard Metals Based on Cemented Borides, in Lectures on Advanced Ceramics, Uchida Rokakuho, Tokyo

17. K. Portnoi, G. Samsonov, 1960 Gurnal Prikladnoi Chimii,33 577584

18. M. Schouler, M. Ducarroir, C. Bernard, 1983 Rev.Int.Hautes Temp. Refract. 20 2630

19. V. Solozhenko, 2001 Diamond Relat.Mater. N10, 22282234

20. V. Solozhenko, E. Gregoryanz, 2005 Synthesis superhard materials, Materials Today, Elsevier Ltd.,7685

21. H. Stadelmaier, J. Lowder, 1967 Metall (Berlin),N21 (10),10231102

22. Z. Zachariev, 2001 "Neue Superharte Kompositionswerkstoffe", Metall, (Internationale Fachzeitschrift fur Metallurgie), Giesel Verlag GmbH, Isernhagen, 2387

23. Z. Zakhariev, D. Radev, 1988 Properties of polycristaline boron carbide sintered in the presence of W2B5 with out pressing, J.Materials Science Letters, 7 695697

24. Z. Zakhariev, M. Ivanova, T. Serebriakova, 1993 Hard Materials Based on Cemented TiB2 -WB-Co Alloys, XI Inter.Symp.Boron, Borides and Related Compounds, Tsukuba, Japan

25. Z. Zakhariev, M. Marinov, R. Zlateva, Ch. Chistov, 1987 A new combination of coatings on carbide cutting tools, Surface and Coatings Technology, 31 265273

26. Z. Zakhariev, R. Ziateva, K. Petrov, 1986 Microhardness and high-temperature oxidation stability of CoWB, J. Less-Common Metals, 117 129133

27. Z. Zakhariev, N. Belopitov, N. Razkazov, 1970 Pat.Bulg. N16115

CITATION

CHAPTER 1

Shilko Serge (2011). Adaptive Composite Materials: Bionics Principles, Abnormal Elasticity, Moving Interfaces, Advances in Composite Materials - Analysis of Natural and Man-Made Materials, Dr. Pavla Tesinova (Ed.), ISBN: 978-953-307-449-8, InTech, DOI: 10.5772/18190.

CHAPTER 2

E. Dado, E.A.B. Koenders and D.B.F. Carvalho (2012). Netcentric Virtual Laboratories for Composite Materials, Composites and Their Properties, Prof. Ning Hu (Ed.), ISBN: 978-953-51-0711-8, InTech, DOI: 10.5772/48705.

CHAPTER 3

Konstantin N. Rozanov, Marina Y. Koledintseva and Eugene P. Yelsukov (2012). Frequency-Dependent Effective Material Parameters of Composites as a Function of Inclusion Shape, Composites and Their Properties, Prof. Ning Hu (Ed.), ISBN: 978-953-51-0711-8, InTech, DOI: 10.5772/48769.

CHAPTER 4

R. Caputo, L. De Sio, A. Veltri, A. V. Sukhov, N. V. Tabiryan and C. P.Umeton (2011). POLICRYPS Composite Materials: Features and Applications, Advances in Composite Materials - Analysis of Natural and Man-Made Materials, Dr. Pavla Tesinova (Ed.), ISBN: 978-953-307-449-8, InTech, DOI: 10.5772/17137.

CHAPTER 5

Paciornik, S., & d'Almeida, J.. (2010). Digital microscopy and image analysis applied to composite materials characterization. Matéria (Rio de Janeiro), 15(2), 172-181. Retrieved March 12, 2016, from http://www.scielo.br/scielo.php?script=sci_arttext&pid=S1517-70762010000200013&lng=en&tlng=en.

CHAPTER 6

Yong X. Gan, Effect of Interface Structure on Mechanical Properties of Advanced Composite Materials, doi: 10.3390/ijms1012511

CHAPTER 7

B.L. Sharma and Parshotam Lal (2011). Growth Reinforcing Composite Materials from Liquid Phase: Mechanical and Microstructural Parameters Relationship Essentially Evincing the Predominance of an Akin Mass Composition over the Domain of Compositions, Metal, Ceramic and Polymeric Composites for Various Uses, Dr. John Cuppoletti (Ed.), ISBN: 978-953-307-353-8, InTech, DOI: 10.5772/18216.

CHAPTER 8

Ali Emamian, Stephen F. Corbin and Amir Khajepour (2011). In-Situ Deposition of Metal Matrix Composite in Fe-Ti-C System Using Laser Cladding Process, Metal, Ceramic and Polymeric Composites for Various Uses, Dr. John Cuppoletti (Ed.), ISBN: 978-953-307-353-8, InTech, DOI: 10.5772/10593.

CHAPTER 9

Chien-Lin Huang, Ching-Wen Lou, Chi-Fan Liu, Chen-Hung Huang, Xiao-Min Song, and Jia-Horng Lin Polypropylene/Graphene and Polypropylene/Carbon Fiber Conductive Composites: Mechanical, Crystallization and Electromagnetic Properties, doi:10.3390/app5041196

CHAPTER 10

Zakhariev (2011). New Superhard Ternary Borides in Composite Materials, Metal, Ceramic and Polymeric Composites for Various Uses, Dr. John Cuppoletti (Ed.), ISBN: 978-953-307-353-8, InTech, DOI: 10.5772/21651.

INDEX